APPLIED VIRTUALITY
BOOK SERIES
**PRINTED PHYSICS
—METALITHIKUM I**
EDITED BY
VERA BÜHLMANN,
LUDGER HOVESTADT

AMBRA |V

EDITORS
PROF. DR. LUDGER HOVESTADT: Chair for Computer Aided Architectural Design CAAD, Institute for Technology in Architecture ITA, Swiss Federal Institute of Technology ETH in Zürich, Switzerland.
DR. VERA BÜHLMANN: Laboratory for Applied Virtuality at the Chair for Computer Aided Architectural Design CAAD, Institute for Technology in Architecture ITA, Swiss Federal Institute of Technology ETH in Zürich, Switzerland.

The Metalithikum Books are part of the Applied Virtuality Book Series, edited by Ludger Hovestadt and Vera Bühlmann (Swiss Federal Institute of Technology ETH in Zürich, Switzerland).
www.caad.ethz.ch

This work is subject to copyright.
All rights are reserved, whether the whole or part of the material is concerned, specifically those of translation, reprinting, re-use of illustrations, broadcasting, reproduction by photocopying machines or similar means, and storage in data banks.

© 2013 AMBRA | V
AMBRA | V is part of Medecco Holding GmbH, Vienna
Printed in Germany

PRODUCT LIABILITY: The publisher can give no guarantee for the information contained in this book. The use of registered names, trademarks, etc. in this publication does not imply, even in the absence of a specific statement, that such names are exempt from the relevant protective laws and regulations and are therefore free for general use.

The publisher and editor kindly wish to inform you that in some cases, despite efforts to do so, the obtaining of copyright permissions and usage of excerpts of text is not always successful.

LAYOUT AND COVER DESIGN: onlab, Nicolas Bourquin, Thibaud Tissot, Niloufar Tajeri, Pernilla Forsberg, Johanna Klein, D-Berlin, www.onlab.ch
TYPEFACE: Korpus, binnenland (www.binnenland.ch)
TRANSLATION: Erik Larsen and Kirsten Leuschner (Bringing and Positioning: Ways of technology?); Reinhart R. Fischer (A Fantastic Genealogy of the Printable; Primary Abundance—Urban Philosophy; That Centre-Point Thing; Digital Cathedrals).
IMAGE REFERENCES (ASSISTANCE): Mario Guala, Aoife Rosenmeyer
PRINTING AND BINDING: Strauss GmbH, D-Mörlenbach

Printed on acid-free and chlorine-free bleached paper

With 62 figures

ISSN 2196-3118

ISBN 978-3-99043-570-0 AMBRA | V

TABLE OF CONTENTS

	ON THE BOOK SERIES	6
	INTRODUCTION — PRINTED PHYSICS	11
I	A FANTASTIC GENEALOGY OF THE PRINTABLE	17
	LUDGER HOVESTADT	

I HELLO SUN 20 · MUCH-PRINTED 20 · BABYLON 21 · THE MYTHICAL SUN 23 · REFLECTING THE COURSE OF THE SUN 24 · PROJECTING INTO REFLECTIONS 24 · THE SHADOW PLANET 25 · WILLIAM'S TRIP TO THE MOON 26 · COMFORT SYSTEMS 27 · RAGE SYSTEMS 29 · HELLO WORLD 30 — II LOSS OF CONTROL 30 · PRINTING WORDS 31 · PRINTING ILLUSTRATIVENESSES 32 · PRINTING ATTRACTIONS 49 — III IN THE PARADOX MILIEU 50 · OUR LANGUAGE GAME 51 · DETERRITORIALIZATION 51 · RETERRITORIALIZATION 54 · DOUBLE ARTICULATION 56 — IV TECHNICAL ARTICULATIONS 56 · THE IMAGE 56 · TRANSFER 57 · THE LETTER 58 · ROTATION 59 · COLOUR 59 · MOTION 60 · FORM 60 · FAÇADES 61 · PRINTERS 62 · MIRRORS 62 · SOUND 63 · MATERIALS 63 · ENERGY 64 — V ALWAYS ON 65
INDEXICAL MARKINGS OF THE TOPICS DISCUSSED 67

II	TECHNOLOGY AND MODALITY	71
	HANS POSER	

I MODALITIES AND THEIR IMPORTANCE 74 — II TECHNOLOGY AND NECESSITY 79 — III TECHNOLOGY AND POSSIBILITY 87 · TECHNOLOGICAL POSSIBILITY—EPISTEMIC OR ONTOLOGICAL? 88 · ACTUALIZABILITY 89 · ELEMENTARY AND THEORETICAL TECHNOLOGICAL POSSIBILITY 94 · THE POTENTIALITY OF AN ARTEFACT 97 — IV HOW TO DEAL WITH CONTINGENCY 98 — V EPISTEMIC-TECHNOLOGICAL POSSIBILITY 99 — VI FICTION, REALITY AND ACTUALITY 100 · VIRTUALITY, REALITY AND ACTUALITY 101 · VIRTUALITY AND POSSIBILITY 104 · THINKING IN NEW MODALITIES 107 — VII FIRST RESULTS 110

III	PRIMARY ABUNDANCE, URBAN PHILOSOPHY — INFORMATION AND THE FORM OF ACTUALITY VERA BÜHLMANN	113

I PRELUDE 117 — II ACTUALITY 118 — III CAPACITY 121 — IV ELECTRICITY 126 — V SUN 128 — VI HOUSEHOLDING WITH CULTURE 130 — VII WITHIN THE URBAN 134 — VIII MOTORICS OF SYMBOLS AND ENERGY 136 — IX VALUES 140 — X INVARIANCES 143 — XI MEDIALITY 146 — XII DOUBLE ARTICULATION 148 — XIII CODA 149
INDEXICAL MARKINGS OF THE TOPICS DISCUSSED 150

IV	THAT CENTRE-POINT THING — THE THEORY MODEL IN MODEL THEORY KLAUS WASSERMANN	155

I UP AND IN 159 — II THE DEMISE OF THE MIDDLE (LONG LIVE THE MIDDLE) 161 — III MODEL 169 — IV DE-CENTREMENT OF MODELS 171 — V THE PRACTICE OF THE DE-CENTRED MODEL 175 — VI THE NON-ASCERTAINABLE LOCUS 181
INDEXICAL MARKINGS OF THE TOPICS DISCUSSED 185

V	DIGITAL CATHEDRALS — A FEW REMARKS ON THE QUESTION OF APPROPRIATENESS WITHIN ARCHITECTURE HELMUT GEISERT	189

INDEXICAL MARKINGS OF THE TOPICS DISCUSSED 204

VI	BRINGING AND POSITIONING: WAYS OF TECHNOLOGY? — APPROACHING HEIDEGGER'S THOUGHT ON TECHNOLOGY HANS-DIETER BAHR	211

INDEXICAL MARKINGS OF THE TOPICS DISCUSSED 227

IMAGE REFERENCES   232

THE METALITHICUM BOOKS
VERA BÜHLMANN, LUDGER HOVESTADT

Only one hundred years ago, hardly any scientist of renown would have been unaware of philosophy, and hardly any artist or architect uninformed about up-to-date technology and mathematics. Today, our ability to explain and explicate our own work within a shared horizon of assumptions and values beyond our specific scientific community has, perhaps paradoxically, turned into an inability and resulted to some degree in a kind of speechlessness. Only rarely now is it thought important that we relate our work to, and integrate it with, an overall context that is in itself "on the table" and up for consideration. More and more, that kind of context is taken for granted, without any need for active articulation, refinement or development. At the same time though, the media are full of news stories about catastrophes, crises, and an impending doom that cannot, it seems, be warded off. Climate change, shortage of resources and population growth, urbanization, and this is just naming a few of the critical issues today. Quite obviously, the notion of such an overall context, both implicit and assumed, is extremely strained, if not indeed overstretched today. This all is widely acknowledged—the UNESCO Division of Foresight, Philosophy and Human Sciences in Paris, for example, launched a discourse on this subject in their 21st-Century Talks and Dialogues under the heading *The Future of Values*. The companion book, published in several languages simultaneously in 2004, is structured in three parts, and includes one chapter on the ethical issues of values and nihilism lying ahead, another chapter on technological progress and globalization, as well as a third chapter on the future of science, knowledge, and future studies. What remains strangely implicit, and in that manner ignored here, in a way that is typical of this inarticulacy with regard to an overall context mentioned above, is the societal, scientific and cultural role that inevitably is ascribed to technology against the backdrop of such discussions, along with the expectations that are associated with that role of technology. In the Metalithicum Conferences, we tend to regard technology in an extended sense. Along with its respective solution-oriented application to the sciences, culture, economics and politics, we think that technology needs to be considered more fundamentally, especially regarding the semiotic and mathematical-philosophical aspects it incorporates. From this perspective, we see in technology a common denominator for facilitating a discourse that seems to have been largely lost from today's discursive landscape, the degree of its disappearance inversely proportional to the increasingly central role technology plays in every domain of our lives. To us, such a discourse seems crucial if we are to develop adequate schemes for thinking through the potentials of today's technology, something that is in turn essential for all planning. Our stance is an architectural and, in the philosophical sense, an architectonic one. Our main interest centres on the potentials of information technology, and how we can get used to the utterly changed infrastructures they have brought us.

But have our infrastructures really changed substantially? Or is it merely the case that a new level of media networks have emerged on top of technology with which we are already familiar? Are the "new" and digital media simply populating and exploiting, in a parasitic sense, the capacities of modern industrial infrastructures that have brought prosperity and wealth to so many? In his contribution to the UNESCO dialogues, Paul Kennedy was still convinced: "In the Arabic world, 3% of the population has access to the internet. In Africa, it's less than 1%. This situation won't improve as long as the infrastructures remain in their current state. It won't change, as long as these countries lack electrification, telephone wiring and telephones, and as long as the people there can't afford either computers or software. If knowledge is indeed power, then the developing countries today are more powerless than they were thirty years ago, before the advent of the internet." Our experience since then has allowed us to see things a little differently. The "Arab Spring" that brought simultaneous political revolutions in several Arabic countries from the end of 2010 and in the first half of 2011, in a development that, at the time of writing, still continues, gives credence and facticity to the cultural impact of digital media, to a degree that was unexpected or previously deemed improbable by many. Meanwhile in India, by May 2011, an astounding 200 million people had been recorded as owning a mobile phone, even though they are living without electricity in their homes, or even in their villages (some 600 million people are still living without electricity in India), and need to take quite some trouble to provide power for them every once in a while. Every month, another 20 million become mobile phone users, in India alone. Here, information technology has achieved what no administration, no mechanical infrastructure, no research and no aid has been capable of: enabling people in developing areas of the world to use standard, state-of-the-art, technological infrastructures, for their own benefit.

We read these events as strong indicators for just how limited the applicability of our noetic schemes is for thinking through long-term developments. These schemes have evolved from our experience of prosperity in times of strong modern nation states and industrial technology with matching economics. They go along with notions of centredness for thinking about control, notions of linearity and nested recursion, of processes and grids, and of mechanical patterns of cause and effect used for planning. It is a truism, perhaps, to point out that these notions do not fit information technology very well. They are stressed and overstrained by the volatile associativity that emerges from logistic networks and disperses throughout user populations. Going by our inherited notions, industrial infrastructures appear to be used as a playground for what is called, somewhat helplessly, "consumer culture" or "the culture industry". But in the case of India, for example, what came back as a result of the success of mobile telephony, astonishingly, were new infrastructural solutions. With no banks and no cash machines on hand, people simply invented the means to transfer money and pay by SMS. Yet the standards developed for micro-banking today can be referred to and linked up with solutions that exist for other areas, such as energy provision maintained by photovoltaics and micro-grids, for example. But this is not the place to present scenarios. We would simply like to invite you to consider the profound extent to which codes, protocols, or algorithms, standards such as ASCII, barcodes, MP3, or the Google and Facebook algorithms, have challenged our established economic, political and cultural infrastructures. From this we get a sense of the potentials that come with information technology, directly proportional to these challenges. We deliberately call them potentials, because we are interested in developing adequate noetic schemes for integrating them into thinking about information technology from an infrastructural perspective. We are interested in how these potentials and dynamics can be applied to finding ways of dealing with the great topics of our time.

In another contribution to the UNESCO dialogues mentioned above, Michel Serres observed, somewhat emphatically: "Today's science has nothing to do with the science that existed just a few decades ago." Computers and IT bring us the tools for statistical modelling, simulation and visualization techniques, and an immense increase in accessibility of data and literature beyond disciplinary boundaries. With the conferences that are documented in the Applied Virtuality book series, of which this is the first volume, our main interest lies in how to gain a methodological apparatus for using the potentials and dynamics that are specific to information technology and applying them to dealing with the global challenges that are characteristic of our times, by referring them to a notion of reality we assume will never be "fully" understood.

The prerequisite for making this possible is a regard for, and estimation of, the power of invention, abstraction and symbolization that we have been able to apply, in past centuries and millennia, in order to come up with ever evolving ways of looking at nature, cities, at trade and exchange, at knowledge and politics, the cosmos and matter, and increasingly reflected, at our ways of looking, speaking, representing. Rooted in their respective historical cognitive frames of reference, we have been able to find ever new solutions for existential challenges. There has most likely never been any such thing as a prototype for coordinate systems: their detachment from substance-space and its formal symbolization result from acts of abstraction. Plato may have already considered the idea of a vacuum, yet he thought it "inconceivable"; nevertheless, this notion of the vacuum inspired abstract thought for ages, before Otto von Guericke invented the first vacuum pump as a technological device in 1654. Electricity was thought of as sent by the gods in thunderstorms before the algebraic mathematics of imaginary and complex numbers were developed along with the structures that allowed us to domesticate it. Today, we imagine the atomic structure of matter by means of orbital models gained from a better understanding of electricity.

So, in short, we do not share the idea that characterizing our time as post-anything is very helpful. While we agree that we seem to be somewhat stuck within certain mindsets today, we do not consider it at all plausible that any kind of concept or model, political or otherwise, will ever come close to anything resembling a natural and objective closure. The concepts behind any assumption of an End to History—whether this be in the Hegelian, the Marxian, or the more recent Fukuyama sense—stem from the 19th century, when Europe was at its peak in terms of imperialist expansion. To resurrect them today, in the light of our demographic, climatic and resources-related problems, to us seems a romantically dangerous thing to do.

By now it is safe to say that technology is not simply technology, but has changed character over time, perhaps even, as Martin Heidegger put it, it has changed "modalities in its essence". In order to reflect this spectrum, we propose to engage with a twin story, which we postulate has always accompanied our technical evolution. Historically, the evolution of technics is commonly associated with the anthropological era called the Neolithic revolution, which marks the emergence of early settlements. We suggest calling our twin story Metalithicum. As the very means by which we have been able to articulate our historical accounts, metalithic technics has always accompanied Neolithic technics, yet in its symbolic character as both means and medium it has remained largely invisible. The Metalithicum is ill suited for apostles of a new origin, nor is it a utopian projection of times to come. Rather, we wish to see in it a stance for engaging with the historicity of our culture. As such, it might help to bring onto the stage as a theme of its own an empirical approach to the symbolics of the forms and schemes that humans have always applied for the purpose of making sense. This certainly is what drives our interest in the Metalithicum Conferences, which we organize twice a year in a concentrated, semi-public setting. As participants, we invite people from very different backgrounds—architects and engineers, human and natural scientists, scholars of humanities, historians—or, to put it more generally and simply, people who are interested in better understanding the wide cultural implications and potentials of contemporary technology. This characterizes the audience for whom this book is written as well.

We are very grateful for the opportunity of collaborating with the Werner Oechslin Library Foundation in Einsiedeln. The Library chiefly assembles source texts on architectural theory and related areas in original editions, extending from the 15th to the 20th century. Over 50,000 volumes document the development of theory and systematic attempts at comprehension and validation in the context of humanities and science. The core area of architecture is augmented, with stringent consistency, by related fields, ranging from art theory to cultural history, and from philosophy to mathematics. Thanks to the extraordinary range and completeness of relevant source texts and the academic and cultural projects based on them, the library is able to provide a comprehensive cultural history perspective. When we first talked to Werner Oechslin about the issue that troubled us most—the lost role of Euclidean geometry for our conceptions of knowledge, and the as yet philosophically unresolved concepts of imaginary and complex numbers and their algebraic modelling spaces—he immediately sensed an opportunity to pursue his passionate interest in what he calls "mental chin-ups" as a form of "mental workout", if not some kind of "thought acrobatics". We would like to express our thanks to him and his team for being such wonderful hosts.

INTRODUCTION
—PRINTED PHYSICS
VERA BÜHLMANN, LUDGER HOVESTADT

The topic Printed Physics takes as its starting point the phenomena observed in recent developments in information technology, by which materials can have their physical characteristics formally analysed, technologically constructed and (bio-)chemically synthesized on a symbolic level, and—hence the wording of the title—can be produced industrially, using printing technologies. This manipulation of materials, specifically upgrading them so they become capable of information-technological programming and control functions, is called "doping". Doped materials can be manufactured using a process that bears striking similarities to the printing technologies we are familiar with from the past. The manufacture of digital processors and memory chips for example is in fact reminiscent of lithography and copper etching, and the chemical printing of photographs, and thus comes to continue a line of earlier forms of analogue relief printing methods. In the case of printable solar cells, it can be said that instead of ink on paper, ions are literally being "imprinted" on silicon. Yet there is one important difference, which becomes apparent in the respective notions of "imprinting" and "doping". Unlike any other print product we have known before, this new printed matter plays a genuinely operational role, rather than a primarily descriptive or representational one. What we call printed physics actually refers to tiny electronic devices, produced and distributed on an industrial scale in processes that are akin to those used in the printing and distribution of newspapers.

From a philosophical perspective, there is something interesting that happens in this printing of doped materials. The rationally defined grid, which has been crucial to us for deducing all our physical descriptions of nature, serves here as a frame for projecting the fantastical. Could it not be, is the question we would like to pose with this book, that we are witnessing a development in the field of physical thinking similar to the one that occurred in geometry in the early 19th century? Are we seeing the appearance of non-naturally determined physics as a complement, as it were, to non-Euclidean, projective, algebraic geometry?

As a context for our discussions in the Printed Physics Conference, we suggested a thought experiment: suppose a new enlightenment of physicalist and naturalized rationality and logic were to be announced, brought about and carried by the qualitative and quantitative impacts of the doping of materials and their production through print. How could this be argued?

Polymer electronics printed onto a polyester film in several layers.

A THOUGHT EXPERIMENT   For the purpose of this exercise, let us regard the pre-modern monasteries as proper "production plants" for the copying of the holy scriptures. The subsequent secularization of this process was brought about and carried by the qualitative and quantitative impact of printing, exemplified by the availability of "text" as a medium, the standardization of format and the freedom, for wide sections of society, to access what was once a ritual and sacred description of the world and to take an experimental approach to it. Astonishingly enough, in a reverse analogy, today's factories, businesses and bureaucracies, with their modern industrialization of secularized rationalism, appear like "monasteries" for the copying of physical "regularities". Systemically integrated into orders of higher and lesser function, and cybernetically implemented in multifarious information-technological infrastructures, these law-like regularities act as impulse generators for societal, scientific and economic processes. Our thought experiment suggests to test putting "functional infrastructures" in the place of the "holy scripture" in the pre-Gutenberg era, before the rise of modern, experimental science.

The question that we would like to propose for consideration, not so much as a scientific or philosophical hypothesis than as our thought experiment, arises from this and goes as follows: if the printing press promoted the secularization of mental horizons in philosophy and modern science, is it not possible that these new printing technologies could bring about a further secularization; the secularization of a naturalized rationality principle? There are plenty of indications to suggest that contemporary production methods assert the same principles as were used back in the days of Gutenberg, but on a new plateau. We want to consider this as a plausible scenario by following two lines of argument, one qualitative and one quantitative.

THE QUALITATIVE ARGUMENT It is neither a new nor a bold thesis, at this point, to posit that information technology is essentially concerned with symbolic operations, and that these symbolic operations—not only from a philosophical but also from a technological perspective—cannot appropriately be reduced to the causal connections that are formulated in physics. Information is information, not matter or energy, as Norbert Wiener suggested more than half a century ago.[1]
The effectiveness of information technology does not develop on the same level as the effectiveness of heat, levers, gears or any other mechanical device. Information technology controls the physical conditions symbolically. This means that information technology is operating on a different "substrate" from the physical-material technology of clearly identifiable cause and effect, its symbolicity turns it into a "medial" substrate—medial in the sense that it allows for different possible ways of operation.
A seemingly natural objection that might be raised at this point would perhaps be to regard the electric current as a type of "physicality", and in so doing lend a familiar substratum in the traditional vein to the "void"of the symbolic. This does not, however, solve the problem of defining the relationship between electricity and symbols: quite the opposite. To this day, there is no coherent proposal as to how we are to view electric currents from a physical perspective: should we regard its elements as fields, as waves, as particles or as impulses? The situation is no simpler from a philosophical perspective. All electronic technology is based on precisely that kind of algebraic analysis of symbolic operations which not only triggered, due to their genuine un-intuitivity and non-representationality, but also exacerbated what is known today as the foundation crisis of the sciences, around the turn of the 20th century. Nevertheless, we experience electric current on a daily basis as the ubiquitous availability of energy, as the potential for potential, so to speak. If one were to understand this availability not merely as a

---

[1] Norbert Wiener, *Cybernetics*. MIT Press, Massachusetts 1948, p. 155.

phenomenal characteristic that has emerged as an afterthought, but as a constitutive element of electricity, a rift would open up in the relationship between symbolization and physics that is in no way inferior to the rift that exists between logical geometry a priori and geometric description of reality a posteriori.

THE QUANTITATIVE ARGUMENT   If, since its invention, information technology has primarily been used to refine the regulating, switching and controlling operations of mechanical equipment, especially equipment that uses electric current, today a completely different dimension of application is being defined. With electronic and information-technological processes, materials can be modified and synthesized not only in their qualitative properties, but also in their physical behaviour, which is to say in their temporal energetic constitution. Artificial materials can now be produced by printing a synthesized ion and semiconductor structure (this is how silicon-based semiconductors are made, for example, but the principle is the same for organic carriers), and in combinations that are not familiar to us from nature. Photovoltaics are an example of this. They allow us to obtain energy directly from light without any combustion process and also without the interposition of other kinetic or dynamic transformers. It really is possible here to speak of "symbolic physics", not only because the "mechanics" are genuinely symbolically constructed (and not the inverse, which would have the symbolic structure derived from a "natural" mechanical context), but also on the basis of their industrial production processes, which today are rooted in information technology.

The principles of photovoltaics have been known for more than a century and yet they have only recently become a relevant component in the discussion about energy supply. It is only in the last few years that manufacturing techniques have become available that make the production of such programmed materials feasible on a large, industrial scale. It is now possible to produce them in printing processes that spew out physically functional apparatus, just like sheets of newsprint that come off the press at the *New York Times*. This goes hand in hand with a quantitative pricing and production development that is characteristic of information technology and is known as Moore's Law: a doubling of the total amount of produced instances every 18 months, which results in a cost reduction of 30% per year.

The quantitative line of argument assumes that every development requires a critical mass number of normalized instances in order to prevail. Such a large-scale fertile ground for these new dimensions of application in information technology has existed for just a few years now. Even so, with "smart" computer chips, this technology has rapidly established itself as an omnipresent feature of our living environment. These chips potentially allow all electrical devices to behave as components of variably configurable systems. The quantitative distribution of systems-capable entities, in fact symbolic-capable actants, seems in a more serious way than just metaphorically comparable to its antecedent, the modern printing revolution.

This comparison may at first glance seem somewhat exaggerated, but as in the wake of the printing press, we too have experienced several abrupt advances that could not have been foreseen: for example, it took only 10 years for 5 out of 7 billion people—approximately 70% of the current world population—to have potential access to wireless connections via information technology. Today we can, at least potentially, phone one out of two people in the world, irrespective of where on the planet they happen to be at that particular moment. Setting up and establishing the infrastructure needed for mobile telephony took only a decade and yet it already feels commonplace to us today. If we want to assess the meaning of this quantitative argument adequately, we must keep in mind that this same technology would be nearly meaningless, and not just in the case of mobile telephony, if there were only a few thousand instances of it throughout the world. What makes it meaningful is that it has reached critical mass, and rapidly. And in this case too, the technology's fast and wide propagation has its foundation in the exact same production and manufacturing processes: information-technological printing technology. The same structural principle applies to the propagation of TV screens as much as it does to internet access, global positioning systems and the efficacy of Google: what would happen if Google could "only" link 1 million sites and only had 10,000 users that sporadically used its search engines? It would be virtually insignificant. Instead, it has now achieved a level of standardization that no longer just renders some qualities or aspects of our practices or behaviour less meaningful—a side effect of any standardization—but rather tips us "over the edge" into a situation where we are now developing new qualities on the very basis of this quantitative standardization.

THE CONTENTS OF THIS BOOK We sent this thought experiment along with our invitation to the speakers of the Printed Physics Conference in early summer 2010. Their contributions, however, represent their independently formulated positions, and only indirectly refer to our overall theme, mainly in the discussions that followed their lectures. In this book, we print the manuscripts of these lectures as distinct chapters, and add brief summaries of the main lines of argument, as followed up in the discussions afterwards. These summaries take an indexing character; they are meant to provide a kind of conceptual mapping of the thematic landscapes through which we wandered, highlighting the most important topical reference points that were raised. In the first chapter, "A Fantastic Genealogy of the Printable", Ludger Hovestadt presents the current innovations in electronic engineering devices, available today for architectural application and integration, on all levels from design to construction and planning. Furthermore, he provides, in a historical account of what he calls "serious storytelling", a conceptual model for considering the specific potency of digital technology. The second chapter, "Technology and Modality", presents an article by Hans Poser who, as we should point out, was unable to attend the first conference in person, but has kindly allowed us to publish his article in our book. In it, he provides strong arguments as to why philosophy needs to pay attention to state-of-the-art technology when offering notions of reality and, directly related to that, notions of possibility and feasibility. In the third chapter, "Primary Abundance, Urban Philosophy—Information and the Form of Actuality", Vera Bühlmann suggests taking a capacity and capability oriented view of the study of information, and reflects on the intimate and co-constitutive relation between philosophical thought and an idea of citiness; her special emphasis thereby lies on the role of technology in that relationship. The fourth chapter, entitled "That Centre-Point Thing—The Theory Model in Model Theory", investigates the philosophical conditions for a machine-based episteme. Klaus Wassermann argues that we are currently experiencing an historic movement that he calls "de-centrement", and which he demarcates from the more common notions of decentralization and deterritorialization, in that this assumed turn towards ever-increasing de-centrement not only challenges any foundation, rules, structures, procedures and patterns that have served us so far for comprehending the world, but most crucially also urges us to calibrate anew the role of the model itself in order to arrive at a philosophical notion of information. In a fifth chapter entitled "Digital Cathedrals", Helmut Geisert challenges the book's emphasis on relating digital technology to earlier printing technology with a retro-projection of how, throughout the 19th and early 20th centuries, people had reflected on the relative cultural impact of Gutenberg's printing revolution. The article reveals a relationship that is as fundamental as it is troubling, between materialist thought and the problem of how to establish notions of proportionality and appropriateness within architecture that has become secularized and profane. In a final chapter, "Bringing and Positioning: Ways of Technology?" Hans-Dieter Bahr introduces Martin Heidegger's thought on technology, and his postulated change in modern technology's essential way of operating. By characterizing it as a "challenging-forth and ordering", rather than as a "setting and maintaining-in-place", Heidegger had introduced many of the core issues behind the conflict between support and control, which we are striving to come to terms with today with increasing urgency.

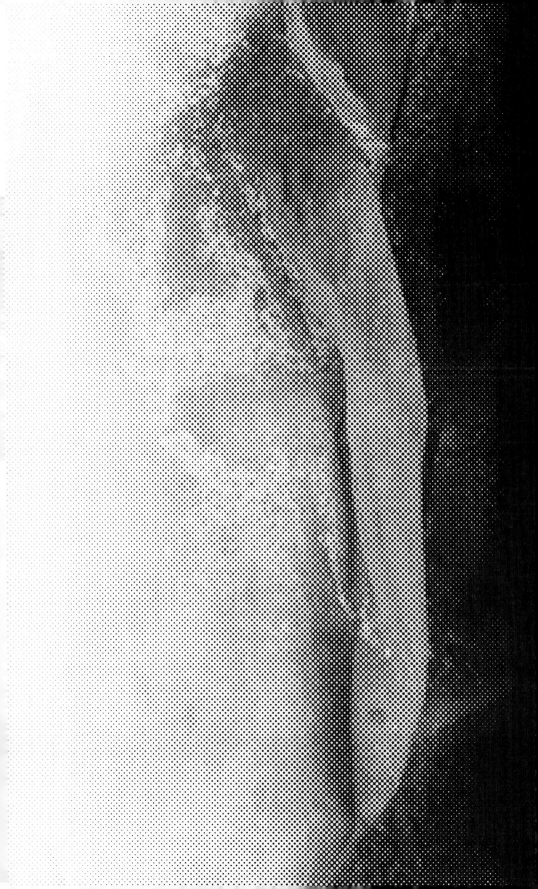

I A FANTASTIC GENEALOGY OF THE PRINTABLE
LUDGER HOVESTADT

I HELLO SUN 20 · MUCH-PRINTED 20 · BABYLON 21 · THE MYTHICAL SUN 23 · REFLECTING THE COURSE OF THE SUN 24 · PROJECTING INTO REFLECTIONS 24 · THE SHADOW PLANET 25 · WILLIAM'S TRIP TO THE MOON 26 · COMFORT SYSTEMS 27 · RAGE SYSTEMS 29 · HELLO WORLD 30 — II LOSS OF CONTROL 30 · PRINTING WORDS 31 · PRINTING ILLUSTRATIVENESSES 32 · PRINTING ATTRACTIONS 49 — III IN THE PARADOX MILIEU 50 · OUR LANGUAGE GAME 51 · DETERRITORIALIZATION 51 · RETERRITORIALIZATION 54 · DOUBLE ARTICULATION 56 — IV TECHNICAL ARTICULATIONS 56 · THE IMAGE 56 · TRANSFER 57 · THE LETTER 58 · ROTATION 59 · COLOUR 59 · MOTION 60 · FORM 60 · FAÇADES 61 · PRINTERS 62 · MIRRORS 62 · SOUND 63 · MATERIALS 63 · ENERGY 64 — V ALWAYS ON 65

LUDGER HOVESTADT is professor for Computer Aided Architectural Design at the Institute for Information Technology in Architecture ITA at the Swiss Federal Institute for Technology (ETH) Zurich. He leads an interdisciplinary team of researchers who understand themselves as technology scouts for architecture. His core interest is in the design of infrastructures, which he defines from an application perspective according to the possibility spaces they are able to provide. He has founded several spin-offs in the fields of digital fabrication, building information models, and building automation. His recent publications include: Beyond the Grid. Architecture and Information Technology. Applications of a Digital Architectonic. Birkhäuser Basel / Boston 2009.

This text is fast, and a tad impatient. Not because it is unable to await the appearance of a foreseeable idea, nor does it indeed at all mean to formulate such an idea in the concrete. No, this text is somewhat impatient with the manner in which, in the few contexts where technology is culture-historically discussed at all, very fundamental shifts are being ignored. Or at least excluded from discourse. Information technology is printing technology. Once conceived, it is reproducible at leisure, like a newspaper. For this reason it grows omnipresent so quickly, unlike any mechanical system. And for this reason, we notice a rhizomic net that gets rapidly tighter across the world, in which information—whatever that may be—dashes as electromagnetic phenomena around our planet at close to light-speed. A net that is the substrate not only for instabilities, but likewise for new stabilities that have increasingly begun to replace old order systems. Whereas our forbears cultivated the land under the rhythms of the sun, and along the sun's reflections built cities whose order was thought to be divine, today we are increasingly relinquishing our familiar and secure territories and must learn to articulate what we find valuable.

I   HELLO SUN

MUCH-PRINTED   A lot is being printed today, and fast. Newspapers, periodicals, magazines, booklets, books, folios, atlases, catalogues, albums, posters, pictures, photographs, maps, tickets. A lot is printed-on today. Paper, cardboard, synthetics, foils, fabrics, leather, glass, metals ... Linotype, intaglio, letterpress, silk-screen, stamps, ink pads, exposure, offset, Xerox, fax, laser, wax, microfilm, lithography, stereolithography ... all those colours, Pantone, CMYK ... poems, diaries, novels, manuals, instructions, drawings, sketches, drafts, constructions, diagrams, lists, compilations, collections, notes, choreographies. Even 3D-printing today, lamps, shoes, prostheses, implants, chairs, prototypes. A lot printed, printed fast. A lot described, a lot written-on. Very large, large, small, very small. Mini, micro, nano. Somewhat surprisingly perhaps: processors, memories, analogue-digital—transformers, sensors, brightness detectors, talk, movement, perception, gestures, contact, danger; emitters of light, sounds, images. Things are not only being described, they are actually being generated through script. Much more radically and directly than up to recently by models, layouts, drafts and constructions. Not only images have learned to walk; so has our writing. Script, analytics, reason, dialectic are being media-ized. Our articulations become real, immediate. In accordance with what we value. What they express is no longer mediated through script confirmable by sole reflection. Our thinking must reinvent its reflecting distance to the world.

The first little program one traditionally learns to write is this:

```
main[ ] {
        printf["hello world"];
}
```

Welcome to the world of articulations in which everything gets media-ized that before was written, demonstrated, and accounted for. Our particular interest is directed at the development of things, artefacts, constellations, and compartments in this world.

BABYLON It is hard to imagine how thousands of years ago people did think. Little has been handed down. It is difficult to undertake archæological excursions bypassing our linguistic-scriptorial way of thinking. Guided by self-referential interest in speech and script, 200 years ago someone hit upon 500,000 old Babylonian clay tablets. Strange signs. A strange script. Strange numerical signs. For better explanation, they are directly set up in tabular equivalence with our decimal system.[1] What else could we do, for a start? But how can this be adequately done, without the nought that did not exist then? Our familiar symbol zero, which was not known in Attic Greece, nor by the Romans, nor in the Middle Ages. I am venturing the thought that without a nought, there were neither "numbers" nor "calculus". There were so-called tabular texts, lists with symbols for cross-attribution of words and numerical words, and there were so-called problem-texts, lists with instructions for manipulating these symbols.

↗ [FIG. 1] P. 33

---

[1] Otto Neugebauer, *The Exact Sciences in Antiquity*. 1957; rpt. Dover, New York 1969, chap. 2. See also Howard Eves, *An Introduction to the History of Mathematics*. Saunders College Publishing, Fort Worth 1992, pp. 44–47.

Out of the discovered 500,000 Babylonian tablets, there are just 300 treating mathematical problems. The symbolics and setups were predominantly used for legal and medical instructions and treatises, over mathematical ones.

Not for me the conjuring-up of myths or promotion of progressive ideologies, but rather the examination of articulatory modes in various contexts. And there are striking similarities between such early tabular texts and tabular texts today. Information technology today is unimaginable without the noetic figure of the table. Research into artificial intelligence, so very popular around the 1980s, distinguished between a knowledge base, which was roughly equivalent to the ancient tabular texts, and rules with forward chains[2] and back chains,[3] which correspond roughly to problem-texts; and then there were distinguished inference machines, with which it was attempted to develop automated problem solvers capable of moving, problem-text—as across the tabular texts. One of the rules looked like this:

> if X croaks and X eats flies, then X is a frog.

Today, linked tables in the form of relational databases have come out ahead. Oracle and SAP are the primary players in this market of data storage and management.

So, the question now arises of how to interpret the clear similarities between the thought pattern of today's information technology, and the ideas and scriptorial peculiarities of a period predating our logic, geometry and philosophy. Certainly not so simple, in terms of the history of progress. And certainly not in terms of an expression of some however-shaped fundamental truth. But let us take our quest for today's printing's possible meaning to some further episodes.

Today, we are familiar with numbers, forms and scripts as means of expression, and like to separate the respective order and knowledge forms rigorously and according to disciplines. Jewish gematria or biblical number symbolism, however, point to the fact that the meanings of numbers, forms and words may not be viewed in such strict separation as those disciplines would have it. In the Middle Ages, for example, the word "calculate" still meant "to re-count, enumerate" and was linked, via its inherent notion of accounting, with story-telling, "giving an account". Numbers and texts lived in close vicinity. So they still do in Gottfried Wilhelm Leibniz' monads, his noetic figure surrounding the notion of integrals, or in the fascination attending his calculating machine in the 17th century. Likewise in the notions propounded in George Boole's *Laws of Thought* in the 19th century, or then in Jacques Derrida's philosophy today... and these are but a very few examples.

[2] e.g. in the programming language OPS5
[3] e.g. in the programming language Prolog

Numbers, forms, and words work—and this can be astonishing only to the engineer that I am—in differentiated and open concert directed at ordering our world.

THE MYTHICAL SUN   From a far-away period 3,500 years back, the Babylonian lists described above came down to us, with their symbols for things, and their instructions of how to deal with things. These symbols were neither signs, numbers, nor designations in the modern sense, nor were instructions equivalent to what today we call functions. Rather, their order seems to have been magically inherent to these things. And such magic seems to flow in particular measure from the sun. The Babylonian gnomon, the Egyptian obelisk, or the Chinese gui biao were, through the shadow that they cast, an immediate index in the world, and referred to a cosmic order. People followed the traces of the shadows over the days and the rhythms of the seasons. They cultivated the magic orders under the sun, so as to be able to harvest their needs. A rural order, stabilized by means of tables and rules. The territory as object, security and reference. The sun as the ritual origin.

↗ [FIG. 13] P. 35

Or let us take the classic Chinese mathematical work Zhou Bi Suan Jing (*The Arithmetical Classic of the Gnomon and the Circular Paths of Heaven*),[4] as formulated in various editions between 1000 and 200 BC. In the very title, the work derives the heavenly movements and the arithmetic from the gnomon, the shadows, and the sun. The Pythagorean numbers, familiar to us Europeans, are presented here in an unusually strict geometric order. And again we see appearing, as in the tables, this disturbing parallel to our time and the beginnings of cybernetics.[5] Europeans, on the other hand, are familiar with an open, figurative representation of the Pythagorean proposition. The few figures are integrated in freely formulated phonetic script, under logic control. The word- and number-related symbols are translated into phonetic characters, straight from the ritual relation with things and with the magic that arises from this ritual relation, as conveyed by the gnomon.[6] The words and numbers are not now in direct symbolic relation to things, they are in direct relation to speech about things. Language gets scripturalized in its sounds, and becomes an independent medium for a person speaking. The order of things is no longer—as especially evident with the gnomon—magically inherent in things, but becomes a matter of skill in dealing with language. It is not the things,

↗ [FIG. 2] P. 33

---

4   Cf. e.g. the explanations of Carl B. Boyer, *A History of Mathematics*. John Wiley & Sons, Inc., 2nd edition 1991.
5   This parallel here reminds me in particular of Concrete Art, e.g. Max Bill's graphic work.
6   Michel Serres, "Gnomon: The Beginnings of Geometry in Greece", in: M. Serres (ed.): *A History of Scientific Thought*. Blackwell, Cambridge, Mass. 1995, pp. 73–123.

I   A FANTASTIC GENEALOGY OF THE PRINTABLE   23 or the speaking person, that are mythical any longer, but indeed the speaking about the things themselves.

REFLECTING THE COURSE OF THE SUN   But how can a text be trusted when its authority is being detached from the familiar things and recognized speakers? The sophists, with their rhetoric, developed rather massive suggestive talents, which became artful, powerful and valuable. That wrought much havoc under the sun when the sophists no longer directly followed the shadows of things. How was a new, adequate notion of order going to be established? The way the speaking about oracles, priests, and rhapsodists, in their quality as personified media, were replaced and scripturalized, thus replacing the regulating course of the shadows of things? Reordered by Plato, operationalized by Aristotle, and applied and put through by Aristotle's disciple, Alexander the Great, thought does not follow shadows any more, but the reflections of sunlight.

↗ [FIG. 3] P. 33

↗ [FIG. 38] P. 43

These are being scripturalized as theoretical geometry and logic, and are now ordering, instead of magic objectiveness, magic text-ness. Texts are reflections about the course of shadows. Philosophy grows together with the striving for notions that might occupy the newly freed spaces, as order has now been detached from magic things. This new order was created within a short span of time by seafaring exiles on the periphery of the large territorial empires of the day. Pocket-shadows on the shifting ground of water. The new order is more adaptable, faster, more pragmatic. Stabilized through the phonetic script, the many tables condense into categories, into hierarchies. Script, logic, geometry originated in urban abundance and abstracted from both territorial and nomadic order principles. The cities and the surrounding rural areas now organize within the logically testable texts.

PROJECTING INTO REFLECTIONS   Let us cut it short here and proceed briskly. With the modern age, the meaning of reflections is being notably shifted. As earlier the shadows relative to reflections, now geometry, logic, words, even notions, turn medial. This liquefaction of traditional orders finds new stability in numbers. Analytic geometry means projecting geometry into numerical space, making it calculable. Numbers acquire the now familiar meaning. They cease to be reflected, scripturally ordered object-ness. Formerly, the number three needed three strokes, and calculating was sticking to the lined-up beads of the abacus. Now there are, projected into the reflections gained from tracing the sun's movement, two-nesses or three-nesses. There is not just one order now; there are numerous projections for arriving at a number. Calculation as a process gains precedence of interest over result, object form, geometry. Calculus means recounting. Numbers are now being recounted. They are no longer objects of

↗ [FIG. 39] P. 43

contemplation. These projections into reflected shadows correspond, loosely phrased, to what today is called analytics.

Due to recountabilities, there are then—hardly imaginable today—for the first time negative numbers. How is it possible, it was previously thought, to eat five apples when there were only three? In the Middle Ages, the resulting two apples had to be positively reformulated. Now they may be recounted as a negative two-ness. There are secure ways now of accounting for zero-ness. Thus the expanses of the rational number space are opening up, which becomes calculable, recountable, and gives onto the real number space. Now, through integrals and differentials, the real, no longer the natural, number space is being domesticated. And texts are no longer being internalized by contemplation in monasteries but, following number sequences, externalized, produced, and printed by machines in cities.

These urban analytic number projections into the textuarily reflected solar shadows are powerful. Interest calculus, new maps, new astronomic instruments ... America is being conquered because one has become able and intent to follow the morning sun across the unstable sea.[7] Tons of gold flow back to Europe and destabilize its old territorial make-up. Much can be paid that had not been harvested on one's own lands.[8] In the sunlight there is not just contemplating now, but also constructing in many places. Our societies' energy is no longer harvested primarily from the cultivated orders of the fields. In a recounted history of sediments, in the recounting of the encapsulations of solar energy during prehistoric eras in coalfields, a hundred times more energy is being found. Forever there was charcoal, forever people knew it burns, but it was a poisonous substance. Hence one burned wood. And went through real energy crises because of forest stripping. Now charcoal can be re-counted differently, it can be washed. Thanks to analytics, constructible material can be found from a wider multitude of substances for a wider multitude of applications. The role of agricultural fields is being mostly taken over by factories on coalfields that were for a long time inexhaustible. On the strength of analytically projectively tapping energy flows from the earth, much larger cities grow now than formerly from the surpluses of the surrounding, geometrically ordered territories.

THE SHADOW PLANET   How are we dealing with these riches? Let us make another short cut and jump to two amusing episodes from the beginnings of cybernetics. In, somewhat ambiguous, sympathy with my professional colleague Buckminster Fuller, we turn our view to an extraordinarily radical cartographic representation of our planet. Fuller

↗ [FIG. 24] P. 39

---

7   For a historical discussion in respect to design practices cf. in particular: Bernhard Siegert, *Passage des Digitalen*. Brinkmann und Bose, Berlin 2003.
8   Cf. Michel Foucault's introduction to the analysis of wealth, in: *The Order of Things*, Routledge, London 2002, chapter 6. (French edition: *Les mots et les choses. Une archéologie des sciences humaines*. Gallimard, Paris 1966).

managed in 1946 to get a particular folding of our earth patented. He projects the planet—or rather the planet's landmass—onto the Archimedean solid of the cuboctahedron and positions this projection in such a fashion that the landmasses need folding only in two places (in central Asia and in Mexico). In contrast to other projects of his, a certain pragmatism is evident here. On the one hand, Fuller evaluates the planet according to content (the continents, not the oceans are in the centre of his representation—"continent" traces back to the Latin *continens*, for "containing, holding")—and on the other he folds his map in such a manner that the landmasses can be "leafed through" continuously ("continent" includes also the notion of continuity). By thus giving precedence to the continents over the oceans, he achieves an exceptionally equiareal and coherent representation of the landmasses of our earth, while the problems of this type of projection are being exported to the oceans (2/3 of the global surface, after all). In addition, his colour scheme primarily indicates temperature zones and, thereby, territorial settlement areas. As if the above abstraction of the energy needed for settlements from territory did not exist, nor the experience of industrialized urban development. As if there were no modern means of transportation (such as his much-admired aircraft); as if there were none of the new orders, which deal perfectly well with water, nay even air and cosmic space. As if oceans were not of ecological interest, which Fuller otherwise so prominently formulated. Today it is proven—one is tempted to say, naturally—that the oceans' impact on climate is much greater than that of the familiar urban projected landscapes, and that they "contain" much more than the continents. We discover in these maps a strange fascination with the limitations of geometric idealization. Fuller's map reflects the idea that pacing out fields and understanding the courses of shadows might provide sufficient safety for our global ambitions. This idea reminds us strangely of McLuhan's—in today's views somewhat ingenuously formulated—notion of the "global village". A rebirth of primary and simple shadow plays. A clinging to very old notions of security whose geometrically verifiable textuary representation harks back to times anterior to any printing techniques, anterior to any analytics or arithmetic of modern times.

↗ [FIG. 42] P. 44

WILLIAM'S TRIP TO THE MOON   Another example from that time. William Anders, during the 1968 Apollo 8 mission, travelled along the shadow tracks of the sun, which were analysed on earth, and succeeded in reaching the moon. In time for Christmas, he shoots there, with an optical device, the first ever picture of an earthrise. A fantastic demonstration of the power of sunlight analysis, as well as that of the United States. The whole planet, rising as normally the sun every morning—a dramatically self-referential, analytical Christmas tale. Probably to be bettered only by a small gnomon, installed by the subsequent Apollo 17 mission three years later on the moon. That installation's message is clear: back to origins. The gnomon on the moon tells an ongoing, direct progress story of 4,000 years, from the direct sun indices of those Egyptians and Babylonians ignorant of reflections and analyses, and letting things themselves speak and write—all the way to Apollo on the moon. This tale takes the whole planet and its history into uniform projective control. An immediate image of the "global village".

↗ [FIG. 44] P. 45

COMFORT SYSTEMS   These two cybernetic episodes are the more astonishing as in the 19th century the Renaissance order, based upon recountabilities, or rather arithmetics, was being re-relativized. As the analytical paradoxes could not otherwise be ordered any more, numbers now acquired two principal meanings. It was no longer possible to assume just one recountability. Numbers were separated into ordinals and cardinals. This is a translation of the familiar gnomon-based territorial measurement into number spaces; ordinal numbers play the indexing role of the gnomon, and cardinal numbers represent the total number of single elements of a set. Cardinal numbers can be used, among other things, for calculation just like earlier rational numbers; ordinal numbers help to index the unsettling and paradoxical infinity of real numbers that analytically cannot be ordered. Ordinally indexed, even infinities can, just as finite numbers, be treated as cardinal numbers. Thus, the old paradoxes are becoming separated and encapsulated. Or as per our analogy of shadows, reflections of shadows, and projections into shadows: now there is clear differentiation between that part of projections which projects into the shadow reflections (ordinal numbers), and the other part whose projections bypass the shadow reflections (cardinal numbers). Because the cardinal numbers "just behave as though" they were rational numbers—what they do reference, however, is not ordered by sunlight, but according to the arranged orders in the ordinals. So, in the 19th century the reflections of the shadows are largely projectionally illuminated. The limits of analytics shine up, and in their contact with the transfinite these limits themselves become projections.

↗ [FIG. 30] P. 41

I should like to sharpen this perspective a bit further. I think it plausible that, against this backdrop, the philosophic concepts of classicism and German idealism in the 19th century were grasping for an anchor in classic antiquity, in reaction to the transformation of the classic reflections into modern projections and their power.[9] The classic

---

9   I was completely taken aback when reading lately that Athens as late as 1830 was still a small village of just 5,000 inhabitants at the foot of the Acropolis, and that it took the constructions and projections of one Ludwig I of Bavaria for reinventing not just Munich, but first of all Athens as an eminent symbol of antiquity. It is not astonishing that it happened that way. It is astonishing how successfully the idealistic framework-setting clearly worked; for how else could it be that I, a fairly curious and informed Central European, had never got onto this simple fact? Much rather, I had presumed the ongoing, continuous cultural importance of Athens, with some ups and downs of course, but from antiquity to our times.

↗ [FIG. 36] P. 42

reflections—derived from the course of sun shadows—are now being conjured into some idealized framework, in order to curb the power of projects. For projections able to do more than reflections by themselves. They can do more in an eerie way. The notions of scarce resources, equitable distribution, limits of growth, begin here to stabilize into a massive dispositive order.

With this account, I attempt to gain some distance for perceiving these present and global problems as symptoms of more fundamental developments that in my view can clearly be distinguished in the developments of mathematics. Analytics, as presented earlier, increasingly propels to the foreground the old paradoxes couched in representations of some infinite, some nothingness, and relating to the way of dealing with probabilities, possibilities and necessities. Arithmetic as an instrument, and the rational-number space along with it, are overtaxed with stabilizing these paradoxes. Modern symbolic algebra finds new ways of dealing with the paradoxes, but no notional horizon for reflecting such dealing. The complex number space is born as a new and technologically cultivated contrivance. Thanks to it, electricity, quantum physics, information technology may thrive outside and beyond what might have been achievable with classic analytics and arithmetic. We are comprehending these developments as the beginning of the "end of intuitive apperception" (*Ende der Anschauung*), whose central culture-historical role we described in this text as "projection reflected in accordance with the shadows of the sun".

Thus, since the 19th century there have existed, as a symptom of that development, attempts at learning a new way of stabilizing the securities gained from the sun and through the shadows' course, so as to recover surenesses from the projections-related Weltanschauungen. That is the cultural atmosphere which gives rise to notions such as environment, climate, life or nature, that are of such primary importance today. At this point, let me highlight the fact that modern symbolic algebra, and with it the presently important technologies, such as electricity, nuclear physics, or information and communication technologies of the 19th and 20th centuries, project in a way bypassing the sun's shadows—to stick once more to our image. The shadows of the sun begin to sparkle, the very things begin to gleam, they shine. Our familiar reflections are thrown back at us. The sun is now hardly helpful at giving orientation, a difficulty prominent in particle and wave representations of quantum phenomena. The result at times is related to our viewing. We must learn to ask what it is that we want to appear attractive to us, and no longer how we may apperceive what is shown to us. Next to reflecting and projecting, we must learn to articulate.

↗ [FIG. 16] P. 37

This is new, with consequences that are breathtaking, and seems to be comprehended only very hesitatingly—while having been with us for 150 years, and the new technologies having long since been omnipresent and commonplace. And yet it is hardly ever being talked about. Preferably polemized against. Too complacently do we settle in ordinalities and keep thinking that the corresponding cardinalities were rational and natural—in the sense of "ordered by the sun". We very smugly made ourselves at home with a mass of machines, statistics, insurances, bureaucracies, expertocracies, designs, intended for stabilizing, amid the projectional Weltanschauungen, the old securities of sun and shadow paths. In the projectively reflected realm of shades of calculated "growth limits".

RAGE SYSTEMS   In this sheltered environment, traditional analytics, which treats cardinalities as though they might still pass as natural and rational, can be practised relatively unperturbedly. The paradoxes of reflections and geometrical intuition are being ignored,[10] and sprout such blossoms as Fuller's cardboard planet. Great technical success stories are being recounted on this little planet of numbers that perform as rational numbers. So everything would be hunky-dory if we were not continuously being faced with the limits and paradoxes of our actions. Were we just to follow, for example, the thoughts of Laplace or Maxwell, we would realize immediately that analytics, used everywhere, is, by definition, entropic. Repeated, pure, objectless reflecting of light leads to ever finer reflections and in the end to meaningless white noise—so well-known from those worldwide interwrapped statistics, from the information glut, and the insurance and control systems with which bureaucratic and technocratic apparatuses strive to stabilize decomposing structures. In mediæval-like contemplation, and trust in authority, we are moving, within the analytic limits of our planet, endlessly and with increasing progress, and are everywhere observing entropy. The world has grown so narrow and threatening.

↗ [FIG. 10] P. 36

In the end, there is just 0.2% of total solar energy that may be harvested on the few fertile surfaces of the earth. Attempted constraining into hierarchical orders under the shadows make us discover the bottlenecks. The media display the limits of our world in an escalatingly catastrophist way. We are mad at it. We are mad at ourselves. Mad at Katrina, the hurricane. Al Gore sermonizes that "nothing is more terrifying than truth" and grabs a Nobel prize and an Oscar, all in one go, for exhortations to eat less meat, climb stairs, and ride bicycles. And we vow reformation, by analysing ourselves to death, against the One Number concealed in the cipher CO_2. After the motto "now more of the same, only more so".

↗ [FIG. 45] P. 45

---

10   One of the few contributions aiming at discussing different stances towards this topic from a comparative angle can be found in an article by C.F. von Weizsäcker, "Geometrie und Physik", in: D. Mayr, G. Süssmann (eds), *Space, Time, and Mechanics. Basic Structures of a Physical Theory*. D. Reidel Publishing Company, Dordrecht 1983, pp. 39–86.

HELLO WORLD   In my quality as an architect and IT man, I find it hard to understand this blindness. The so very common and familiar electricity, and the technology that makes it possible, are not, in the described sense, illustrative, not explainable through reflections, do not spring from this narrow, too tiny and furious, shade-bound planet—and yet function brilliantly. They managed to come about because our projections escaped from the shadows, because things no longer just reflect but begin to shimmer. And these shimmering products, such as electricity or telephony, are not even being rejected, at that. Quite the contrary, they find more acclaim than all the restraining attempts back towards the hierarchical orders under the shadows. They are everyday, and familiar, and their not being around is, after 150 years' presence with us, simply unimaginable.

↗ [FIG. 11] P. 36

↗ [FIG. 14] P. 37

Let us close these introductory musings with a picture from Peter Greenaway's wonderful film *The Draughtsman's Contract*. The two protagonists sit relaxedly in a park clearing, taking in the shadow-work of the sun. Admiring the natural luxury of resources and the natural order of things. And projecting themselves, in contemplative imitation of the light-rays, as rightful beneficiaries of these riches under the sun. In another picture from the same film, a little boy peeks at us irreverently through the very same projection device in the opposite direction—straight into the sun, as it were. In this direction, and in this attitude, there lies, as we shall see, a way out of life's conundrum on a planet that has grown narrow and furious. No longer looking ideologically, nay meekly upon the earth and the shadows, but with curiosity, and straight up into the sun and into the light. HELLO SUN!

II   LOSS OF CONTROL

Printing means speedily and unproblematically making many things out of one. Generalizing, multiplying, serializing, reproducing. But likewise repeating, abstracting, potentiating.[11] Our interest will not necessarily go to the reproduction of simpleness, adding of onenesses, but rather to the emerging diversity of print products that originates from the concert of printed products, and depends qualitative-constitutively upon the very number of instances. For a comprehensive view of such manifoldness, it is important to put the question about the quality of the quantities in which these print products today are "there". Google's fairly simple indexing mechanism would be comparably insignificant, were it to index a mere 1,000 documents. Of suchlike, there are hundreds, if not thousands, of similar potency. Mobile telephony would lack significance if there were only 1,000

---

11   For a philosophical discussion of the problems around these issues, cf.: Gilles Deleuze, *Difference and Repetition*. Columbia University Press, New York 1994.

subscribers. We are to learn to think differently about quantities and qualities, hence the interest in printing techniques. For we have grown used to how massively the many-printed books are able to change our world, and we are grateful for it. And today not only books are being printed but, as will be shown, energy, light, heat or movement, as well. There is none of that reflective writing about something any more, none of that merely projective designing of models of things—the very things themselves are being written.

PRINTING WORDS   What are the shifts wrought, in the 15th century, by Johannes Gutenberg and his printing machine upon writtenness? Script itself remained essentially unchanged. Words and numbers remained reflections of the world. Writing could not but be as before. The first printing machines mechanized copying, not writing as such.[12] Until then, reproduction, and preservation of knowledge about the world order, was up to monasteries. Monks were the medium of reproduction and maintenance of knowledge about the world order. Text[13] was being contemplatively reproduced, and sometimes improved.[14] With the advent of the printing machine, reproduction changed to an explicit, palpable mechanism. To a product of reflection. So, the printing machine led to the establishment of effective self-referentiality. Because the language of books was an expression of contemplative thinking. In addition, the machines of that era were expressions of that same thinking. Printing machines are devices for reproducing texts. Contemplative-thought machines reproduce written expressions of that same contemplative thinking. Books and tools were trained to walk, in the sense of functioning within this self-referentiality, and became self-contemplating, reflective. Texts acquire authors, tools acquire motion. Tools become machines.

↗ [FIG. 18] P. 38

But let us recount this idea in another way, too. An important fact seems to be that those printing machines in the beginning cost markedly more than one person was able to earn in a lifetime. Consequently, large amounts of money and value had to be provided or generated on behalf of something as yet unknown, and moreover legally and politically risky. This meant that there had to be a large market for the still very costly books. The new machines originated in an environment of trades-based urban production, on behalf of an urban market.

12 Kittler's amusing dramatizations relative to the advent of the typewriter describe here the first technical changes of new writing. Friedrich Kittler, *Aufschreibesysteme 1800/1900*. Wilhelm Fink Verlag, Munich 1985.
13 "Text" in the singular, not as we understand "text" today—under the unquestioned assumption that there are many texts of many different authors.
14 For technical reasons as well, books had to be recopied over and over again, if they were to last for centuries.

↗ [FIG. 56] P. 48

New content and new formats were wanted. Through printing, the city graduates to a catalyst for the new forms of knowledge, and takes the role of the monasteries. Printing is an important building-stone for that secularization of knowledge. Production. Quantities. Repetitions. Perspectives. Wealth and richness.

Georg Flegel's still life displays this new, vigorous independence of objects, which are now part of a reflected, and no longer geometrical and contemplative, order. The apples, the bread, the wine. Every element of the painting is fresh, and characteristic in colour, texture and bodiliness. Individual, spatially detailed and energetically charged, in precisely observed light refractions. New urban energies from the quantitative surpluses of the orders of the surrounding fields, now domesticized by reflection. That is the nature of Gutenberg's printing machines.

PRINTING ILLUSTRATIVENESSES   What happened in the 19th century to these newfangled reflective energies, and things produced in unknown quantities? They grew so diverse and so omnipresent that they are now themselves being domesticized, like the fields were in their contemplative orders under the sun. Not things, but qualities are now being re-counted and produced. Coal is being washed, and large quantities of produced energies are available. Many cities, larger than before, now spring up on coalfields, and no more amidst some territory. The territory converts into planned, multiform landscapes. Rural agricultural activities are carried over into the manifold planned factories. Contemplative copying, and amending of that one text, turns into those countless projected texts. Dramas, novels, adventure stories, thrillers, travelogues, diaries, encyclopædias. Universities are being created, evolution, history, republics, pedagogy, psychoanalysis, electricity... Objects' reflective energy turns into projective energies flowing through substances. Not the moving objects but the flows of energy are now being domesticized. Quite some pressure builds up around physics and philosophy from this metaphysical onslaught of *enérgeia* that many suspect inside the form of electricity. But then the subject is briskly changed to *dynamis*, more easily comprehensible since it is more directly in keeping with the familiar projected course of shadows. For stabilizing its manifold energy flows, the 19th century reaches to its origins for an anchor. Reflects itself. Reflects the sun following the familiar reflexive paths. Turns into projection. Reflectively illuminates the sun's shadows, rendering them apperceptible not by intuition but by projection. In such self-referentialities, old orders may be cultivated without any need of stepping out of them. Projection of the sun into sun's reflections. Otherwise, all order would have been lost.

↗ [FIG. 20] P. 38

And printing has grown different and more varied. No longer is there the mere imprinting of words in logically arranged characters onto paper. There is variety of materials and processes.   TEXT CONTINUES ↗ [P. 49]

FIG. 1
Plimton 322, a table text of Pythagorean numbers (3, 4, 5), (5, 12, 13) and (16, 63, 65), Babylon.

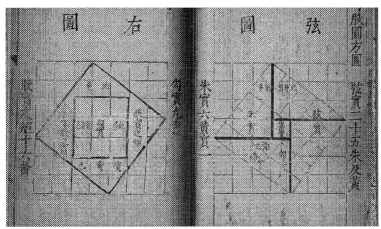

FIG. 2
Page of the Chinese Zhou Bi Suan Jing, 1000–200 BC in strict graphical order.

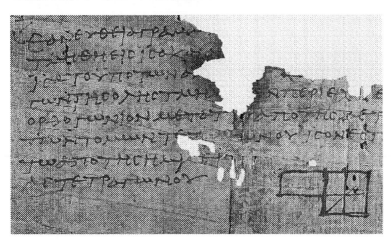

FIG. 3
The oldest extant diagram from Euclid's Elements, vol. II, proposition 5 (c 100 AD). It shows a figure in loose arrangement in the context of phonetic script. In modern algebraic notation, the contents of the diagram are: $ab + (a-b)2/4 = (a+b)2/4$

1 A FANTASTIC GENEALOGY OF THE PRINTABLE

FIG. 4
An Icelandic fisherman in a small village of 300 inhabitants, which owns and can harvest 0.1% of the worlds fish reserve. An incredible wealth per person. Yet, as Andri Snær Magnason points out in his book *Dreamland — a Self-Help Manual for a Frightened Nation* (2008), these people believe themselves to be poor, because "all they've got" is fish.

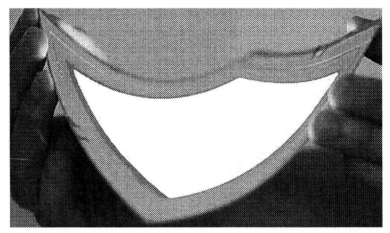

FIG. 5
Doped organic material gives off light when under power.

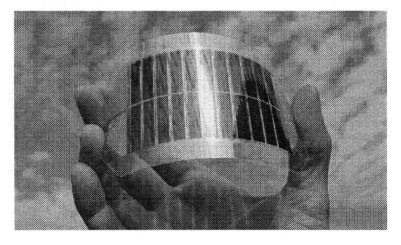

FIG. 6
Doped silicon generates electric current from the solar energy stream.

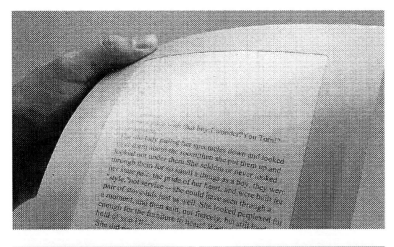

FIG. 7
Paper as each printable page's own printing machine, with newsprint-like contrast, without energy consumption.

FIG. 8
Johannes S. Sistermanns: "inter vue" for violoncello, CD, piezo membrane and sound editing (2004 / 7).

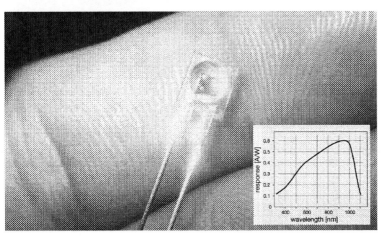

FIG. 9
Photodiode for measuring colour temperature, with curve of sensor characteristic for translating light colour into electric tension.

FIG. 10
Image of a vegetable garden perspectively distorted into a so-called micro-planet, 2008.

FIG. 11
Peter Greenaway's *The Draughtsman's Contract*, 1982.

FIG. 12
Mirror field of the Helvetia building, Herzog & de Meuron, 1989, photo Stefan Tuchila.

FIG. 13
Traditional agriculture, e.g. in Dali, China, cultivation of the orders reflected by of the course of the sun, for obtaining energies.

FIG. 14
Peter Greenaway's
The Draughtsman's Contract, 1982.

FIG. 15
Natural and culturly energy harvest in Willamette Valley, Oregon USA.

FIG. 16
Irrigation of a modern strawberry field as an invisible, analytic projection into the classic order of the sun's path; to me, this is a metaphor for the notion of nature, as it is being applied today.

I A FANTASTIC GENEALOGY OF THE PRINTABLE

FIG. 17
Microscopic image of a printed mirror field, Texas Instruments, 2008.

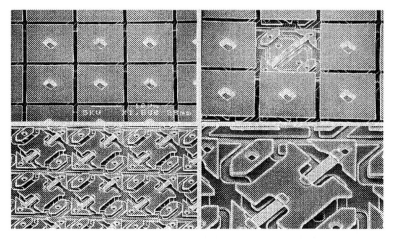

FIG. 18
Lead type.

FIG. 19
Microscopic image of two-colour globules of electronic paper.

FIG. 20
Microscopic picture of a gramophone record.

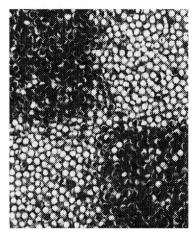
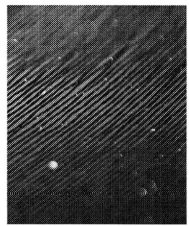

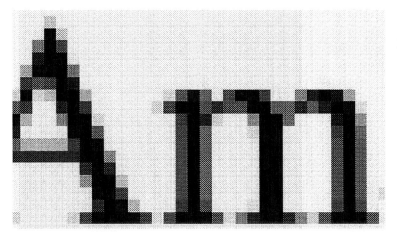

FIG. 21
Enlarged screenshot of letters, Times typeface, anti-aliased.

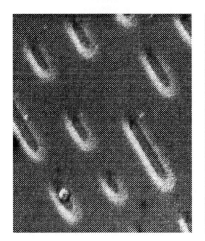 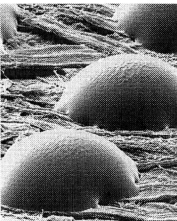

FIG. 22
Microscopic picture of a gramophone record.

FIG. 23
Electron-microscope photograph of inkjet-printer ink droplets on paper.

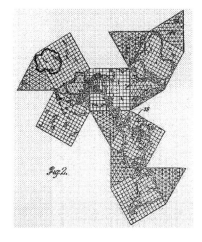 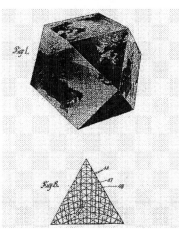

FIG. 24
Patent application for Buckminster Fuller's Dymaxion Map, 1946.

I A FANTASTIC GENEALOGY OF THE PRINTABLE

FIG. 25
Print head of an inkjet printer with a matrix of tiny colour pumps made of piezo crystals.

FIG. 26
Picture of a mirror field for mobile projectors the size of mobile telephones, Texas Instruments, 2008.

FIG. 27
Assembly of a wall by a robot, Gramazio & Kohler, 2008.

FIG. 28
Stepper motor and wiring diagram.

FIG. 29
Experimental ion accelerator, used, e.g. for bombarding silicon crystals with phosphorus atoms.

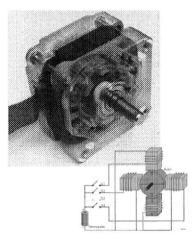

FIG. 30
The sun projected into the sun, ATLAS cavern with magnets for particle acceleration, CERN, 11.2005.

FIG. 31
Production of doped thin-film photovoltaic cells.

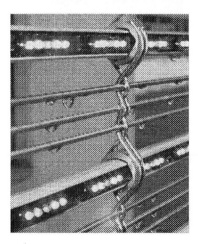

FIG. 32
Array of light-emitting diodes of a transparent media façade, ag4, 2004.

FIG. 33
Piezo crystal as electric Unit of Time.

I A FANTASTIC GENEALOGY OF THE PRINTABLE

FIG. 34
Aura, design by Zaha Hadid, 2008.

FIG. 35
A numerically controlled milling spindle working on a wooden board. The arithmetical, simulated kinematics of the Bézier curve are an example of current, non-Euclidean architecture. Frank Gehry, Düsseldorf 1997–1999.

FIG. 36
Life in secure ordinariness, Greater Kobe, 2008.

FIG. 37
Printed panes on the Brussels Market Square hide construction sites on historic buildings, December 2007.

FIG. 38
Shadows cast by the pyramids of Gizeh, Google Earth. Tracing of the sun's shadows is universally and at all times identically performable; after Michel Serres' illustration regarding the origins of geometry.

FIG. 39
A 19th-century factory is no longer cultivating orders, but the energies of the collieries.

I A FANTASTIC GENEALOGY OF THE PRINTABLE

FIG. 40
Close-up of a flat-screen monitor.

FIG. 41
Processing chamber of an ion accelerator. An ion beam hits material to be doped.

FIG. 42
The first picture of an earthrise, by William Anders, Christmas 1968.

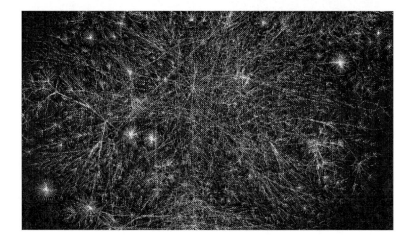

FIG. 43
Map of the Internet—the shining planet.

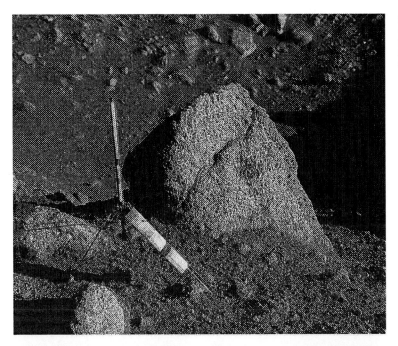

FIG. 44
American gnomon on the moon, Apollo 17, 1971, as stereoscopic image.

FIG. 45
Hurricane Katrina, 2005.

FIG. 46
Light façades in Tokyo.

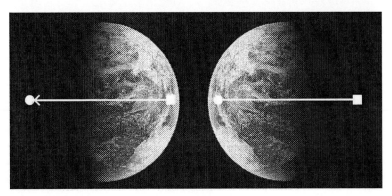

FIG. 47
Networked photovoltaic and light-emitting cells form a culturly rhizome.

FIG. 48
Print mask of the first processor designed as an integrated circuit, Intel 1971.

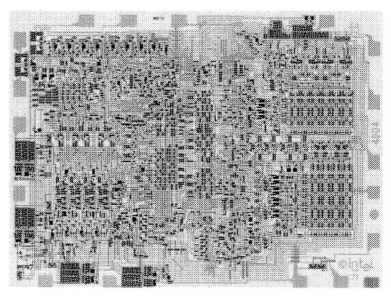

FIG. 49
Functional diagram of a single-slope converter.

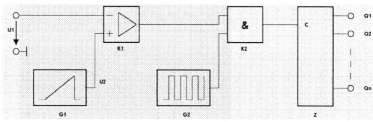

FIG. 50
Electronic setup for driving a matrix, such as a monitor-screen of light-emitting diodes, or, in comparable fashion, e-paper.

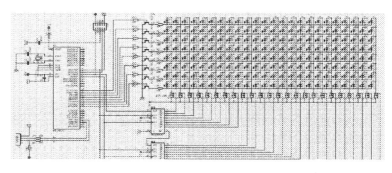

FIG. 51
Column-parallel dural-slope integrating ADC.

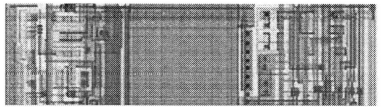

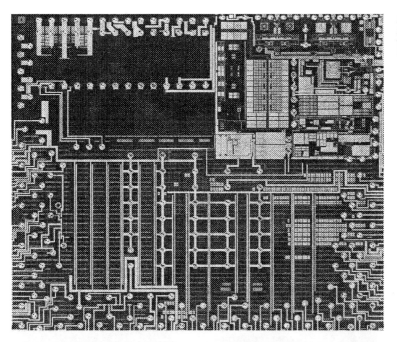

FIG. 52
Single-chip mobile phone, Infineon 2005.

FIG. 53
Microphotograph of an inkjet print; rendering converts the discontinuous colour dots into familiar, continuous appearances.

FIG. 54
The result of code for transporting symbolic points from A to B, here as applied to image transfer.

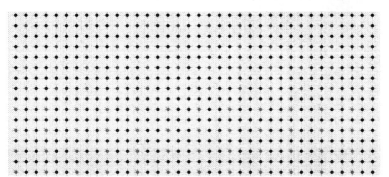

FIG. 55
Diagram of boron (red) and phosphorus (green) doped silicon (black), of a photovoltaic cell.

FIG. 56
Still life with apples, Georg Flegel (1566–1638).

FIG. 57
Apples today—rendered into a logistic assortment.

FIG. 58
Still life, Vincent van Gogh, Amsterdam 1885.

In self-reflective, projected, seemingly intuitive visuality. Such as the gramophone record. What had still been implicitly embodied by characters, between script and speech, becomes now, analytically, explicitly linkable. Sound waves are being analysed as movement, which is then graphically recorded, so that—later perhaps—such form may be reanalysed as movement and retranslated into sound. A direct, intuitive way. Directly printed analytical geometry. Dominance of numbers and arithmetics over notions. Reduction, to mechanical-acoustic transmission, of the spoken word. This seems meagre, or so goes a frequent complaint, compared with the implications of printed, phonetic script. But there will be, as these implications are being further analysed, or so this time promises, additional intuitive, illustrative ways of reproduction. There is no reflection under the sun unsuited for projection. Philosophy and physics, notions and objects lose ground, fade under the stream of projective energies. Chemistry appears with its pragmatic and exceedingly successful symbolizations.

↖ [FIG. 58] P. 48

Much and new stuff can now be printed and reproduced thanks to new motors that might now be called apparatus,[15] and are no longer, similarly to machines, being set in motion, but are in motion. For the first time, every language of the world may be archived and copied without the use of script. Even vocal sound. For the first time, sounds and music of any instrument may be archived and reproduced independently of skill. Colours of sounds, pronunciations, noises are being discovered...

These skills, reductions and potentials happen in parallel to a shift of the notion of language. Saussure's semiology, among others, establishes a new paradigm for semiotics. We think and print whatever we can under the sun. Now constructions, becoming, productions, genealogies, populations, livenesses in van Gogh's still life. What changes in our image of ourselves and our things in the world, what changes to still our stabilities and orders compared to the times of Georg Flegel's still life! New urban attractions from the quantitative affluence of the world's projectively domesticated energies. That's the stuff the new printing products such as gramophone records, photographs, films, comics, novels... are made of.

PRINTING ATTRACTIONS   And today? Today we are busy learning how to cultivate the innumerable projections. Another self-referentiality. We project projectivity. On our data carriers, we bundle single projections: 0 1 0 0 1 1 0 0 1... without the least familiar aspect or intuitivity. Irrelevant, the look of what is printed. No more print image, no more model. We have left the protection of reflective shadows. The very paradoxes around self-referentiality that one tried to avoid by trying so painstakingly to focus the projections into the reflected shadow—they

---

15   Here, we keep following Michel Serres' thoughts from his text "Vorüberlegungen zu einer Theorie allgemeiner Systeme", in: *Hermes IV, Verteilung*. Berlin, Merve 1993, pp. 43–91. Cf. also Vera Bühlmann in this volume, p. 113ff.

K [FIG. 22] P. 39

now become constitutive.[16] 0 1 0 0 1 1 0 0 1... may mean anything. Pure discontinuity, and pure ubiquity. Everything we thought and comprehended analytically, and everything we are yet going to think and comprehend analytically—vaporized and ubiquitous. Without knowing what and how to ask, there is nothing there. Attractive questions open up much wider fields than might analytic ones, beneath the sun's shadows. The sheaving[17] into attractions. Omnipresence of the so far projected turns into the breeding ground for today's acculturating domestication. We must learn to see what we want to see. Then things begin shimmering fantastically. Putting it more mundanely: the tremendous proliferation of print-works and availabilities. Nowadays, more is being printed in one year than humankind ever pronounced. Text, speech, music, sounds, images, films, manuals, programmes. Of main interest today is the way of writing to data, and the reading from data. Perhaps. If only such reading and writing were not in turn itself printed. Not just words, or subsequently measured brightnesses, colour values, or contrasts are being printed. Now—as I shall later describe in more technical detail—even light, heat, movement, energies, or logics are being printed. The potentialities of analytics, cultivation of energies, encapsulated in every single bit. Paradoxes stay within the discontinuities. There is no reflexively derivable context any more. Darkness, except for the energetic gleaming of the networked bit. We move in the movements of others. These movements now make the planet shine. The paradoxes of analytics and of sunlight reflections, so carefully sidetracked by Hegel, Cantor, Hilbert or Saussure. They now turn constitutive. Projections are now undomesticatable within the confines of the shadows. The next 0 or 1 is no longer a matter of reflection, of probing the sunlight for right or wrong, but an uncertain project in abundance of potentials, a dance with analytic paradoxes, with the shimmering light effects we do not know whether to trust. Diversities along the rational reflection space of the sun. A new enlightenment about enlightenment. About which Peirce's algebraic semiotics has more to say then Saussure's semiology. We think and print, what we will-must with the sun. The apples in the supermarket are now individual particularities of some logistics, from universal availability. Renderings.

K [FIG. 57] P. 48

III IN THE PARADOX MILIEU

Why put it so convolutedly, while simpler was on? many will ask. But what now follows comes so close to the stereotype of the usual technical success story, that considerable care had to be put up in

---

16 D.R. Hofstadter gives an introduction to the breadth of the topic in: *Gödel, Escher, Bach: An Eternal Golden Braid*. The Harvester Press, Brighton 1979.
17 Prompted by the mathematical notion of "sheaves" in the context of Grothendieck's algebraic geometry.

order to find a different frame for it. Otherwise, the arguments would unavoidably get side-tracked into the dead-end street of that medially hyperventilating global-crisis hubbub. So we used abstractions as a possible way out of that conundrum.

OUR LANGUAGE GAME   How, therefore, are we to achieve stability in the paradoxical environment outside of the sun's shadows? It appears paradoxical because of its constitutive self-reflexiveness. In following the proposed language game, contemplations about reflections become self-contemplated in the paradox milieu, reflections about projections self-reflected, and projections regarding attraction, in turn, self-projected. We shall link these movements to the different mathematical number spaces. We know natural numbers, which we will link to what we call reflection; rational numbers to what we call projection; and complex numbers to what we call attraction. Put differently, we know the "one text", which will correspond to reflection; we know "every text", corresponding to projection; and "any text", corresponding to what we call attraction. And differently again, printing with type belongs to reflection, with lines to projection, and—as we shall be explicating—with points to attraction.

The dissolving process of self-referential contemplation, reflection and projection we shall call "deterritorialization". The movement inverse to these dissolving processes, will be called "reterritorialization". Through this alternating movement, the geometric and logic, or better yet, epistemic, contemplation order is being self-referentially encapsulated, and these capsules mutate to elements in the construction of new geometric and logic orders. They will be called analytic, or better diastemic, order. With the same movement, the analytic order is being self-referentially and projectively sheaved, and these sheaves turn into articulation elements of new analytic orders. Those, we shall call choreostemic orders. The diastemic order cultivates the differences of encapsulated epistemic orders, the choreostemic order cultivates the differentials of sheaved diastemic orders.[18]

DETERRITORIALIZATION   To begin with, let us turn to the deterritorialization movement. How is it being technically articulated today? On the one hand, we shall examine how the familiar machines of the epistemic order are self-reflectively being encapsulated into motors. Then, we shall investigate how the thus formed, still familiar apparatus of the diastemic order is being self-reflectively sheaved to form sundry applications. Of these distributed and newly bundled motors, we shall speak as of applications of the choreostemic order.

---

18   With this language game, we are pursuing the notion of a generational history of general systems and their technical explications, as proposed and more extensively discussed by Vera Bühlmann in this volume, p. 113ff.

So to the first aspect. A physical quantity, i.e. in our language game an element of the epistemic order, is translated into an electric quantity by a so-called sensor. There are hundreds if not thousands of different sensors, corresponding to various physical quantities, such as pressures, weights, speeds, intensities, frequencies etc. We distinguish between measuring range and measuring accuracy. Sensors are predicated upon unambiguous representability, formability, differentiability, continuity of physical measurement and its representation, projection, deformation etc. This notion of representability puts proportionalities of the epistemic order into a difference, or, put differently, cultivates, as diastemic order, the differences of the epistemic order.

K [FIG. 9] P. 35

The illustration shows a photodiode,[19] which is sensitive to light colour. Today, such diodes are printed in large series and cost just a few pennies. In digital cameras, several millions of them are arrayed into whole fields. These are a specific kind of so-called semiconductors, which convert light colour into electric conductivity. Its characteristic curve shows how electrical conductivity and light colour relate to each other, i.e. there is an arithmetic mathematical transformation of inferring light colour from electrical conductivity. In this case, the calculable assignment of the wavelength between 350 and 1,000 nm is unambiguous. It might be claimed that today all physical quantities can, through sensors, be represented in electric curves, i.e. converted. The notion of continuous transformability is used to conceptualize not only the elements of an apparatus, but their interplay as well, i.e. the apparatus' construction. Machines move within the physical order, whereas apparatus measure and control physical quantities. The difference is like that between a water mill, which we shall call a machine, and the steam engine, which in fact ought to be called a steam apparatus. This exemplifies the difference between technical reflection and technical projection in our language game.

But the electric machines predominant today are not so simply described. Since electric current may originate solely in the choreostemic order of in-symbolic procedures,[20] electric apparatus are constructs within a particularly efficient articulation of the apparatus category: they are constructs of electric articulation. This may be seen from the

---

19 What is a semiconductor, and how is it so easily printable? The discovery that electrical conductivity of materials is not constant along the axes of time and tension, and that this effect is technically exploitable, goes back to Ferdinand Braun (1874). Through inclusion of foreign atoms in a semiconductor—a technical procedure called "doping"—the conductive sensitivity of the semiconductor may be calibrated.

20 For a careful overiew of the backgrounds within field theory, electromagnetism, and the corresponding philosophical challenges in conceiving of this orderability within the symbolic, cf.: Erich Hörl, *Die Heiligen Kanäle. Über die archaische Illusion der Kommunikation*. Diaphanes, Zurich-Berlin 2005, especially the chapters "Die undarstellbare Kommunikation", pp. 93–108, and "Strukturalismus und Feldtheorie", pp. 108–112.

fact that, unlike in the steam apparatus, energy generation is not part of the apparatus itself, or part of some dedicated apparatus, but takes place in whatever fashion, anywhere. Due to this procedural ease vis-à-vis electric "apparaticity", electric cameras, writing apparatus, engines, washing machines etc. are so much smaller, faster, more accurate, and more efficient than purely mechanic apparatus.

In a further step towards choreostemics, the apparatus-like is being sheaved from its distribution. An electrical curve, or mechanical interplay, driving, equilibrating. To this end, a so-called analogue-digital converter transforms an element of the diastemic order into so-called bits. The analogue-digital converter may, just like sensors, be printed or, better, photographically exposed, as an integrated circuit (a pattern of layered, and differently doped semiconductors).

↖ [FIG. 49] P. 46

↖ [FIG. 51] P. 46

↖ [FIG. 48] P. 46

It too is available for just a few cents in unlimited quantities. Let us have a look at the functional diagram of a so-called single-slope converter. It follows from this that an analogue-digital converter functions as follows: a sensor is connected between electric mass and pole 1. A tension U_1 is generated. It corresponds, via the sensor's characteristic curve, to the measured physical quantity (temperature, pressure, brightness, chromaticity, CO_2 etc.). U_1 sensor output be at 3.2 V. A second tension U_2 is being produced via a so-called ramp generator (G_1). 10 times per second it is to go up to 5 V, and back down to 0 V. The electronic circuit K_1 acts as comparator between the two tensions. If the tension of the generator (U_2) is inferior to that of the sensor (U_e), K_1 supplies 5 V, is it superior, K_1 supplies 0 V. Our next functional element is a so-called square-wave generator (G_2) whose tension alternates 1,000 times per second between 0 V and 5 V. The circuit element K_2 will let the U_2 square-waves pass at the precise instant when K_1 supplies a 5 V tension. In our case, this means 64 square-waves. They are being counted by the counter Z, which supplies, on request, the binary-coded number 01000000, which corresponds to 64.

How now to interpret this process? We not only did, as with the electric current and the electric sensors, make energy, and the physical quantities, generally available, but indeed the interplay of these quantities, i.e. the energy flow. Hereby, apparatusness is being made available in generalized form. Because every bit is an apparatus all its own that is being put into whatever kind of relation to other bits. Through simple symbolic algebraic recombination of bits, any apparatus and any machine may now be articulated and construed. Processors, in turn printable patterns of semi-conductors, assume this job technically (more about this later). This is why the computer is often referred to as a universal machine. Putting it succinctly, one might say that, through electric current, energy, as mentioned above, may come anyhow and from anywhere, and that through IT anything may now be done with it.

This thinking goes clearly beyond the apparatus-like idea of constant convertibility, and of the diastemic order. Apparatus conform to physical conditionalities, measure and control physical quantities, are projections towards the territorial, geometric and logic world. The new motors, the applications, on the other hand breed and run apparatus. They follow the idea of enormous availability[21] and combinableness.[22] From energetic and logistic wealth, they articulate apparatus, with specific scarcities and conditionalities.

That's IT potentiality over mechanics. Let us take a blood-pressure meter. As an instrument, it displays my blood pressure in mmHg and compares it to statistic mean values. As an application, it displays the differences against previous measurements, against other body parameters, or against measurements of millions of other people and their individual health profiles, and provides indications as to good or bad experiences of others in similar conditions. On the basis of such applications, marketable and differentiable new value systems may be developed, e.g. new health insurance models. As an articulation issued from logistic wealth. Today, articulations are more powerful than constructions, as may be seen by the way in which computers pushed typewriters off into some niches; as did laser printers to offset printing; Wikipedia to encyclopædias; digital cameras to analogue cameras; Facebook to the photo album; the mobile phone to the landline.

When we say now that today we no longer print characters, nor lines, but dots, these are the underlying developments.

RETERRITORIALIZATION   Summing it up briefly: in deterritorialization, we decomposed an object into differentiable properties. Through sensors, these properties are convertible into electrical properties. In a next step, a single-slope converter digitizes these electrical properties. As a result, any event, any change in physical qualities becomes available over the electronic network, anywhere in the world, in the form of a sequence of ones and noughts. Independently of where and when the respective event happened, and independently of where, when and why interrogation is taking place. This is what is meant, when we describe applications derived from it as articulations from the wealth of digital logistics.

It all sounds rather general and vague. But using Google, one gets an approximate idea of what this wealth of availability might mean. In reality, with every Google search we interrogate the complete explicitly

---

21 The physical substance of this new order consists of the data-technical networks with their meanwhile more than 100 billion processors and 1,000,000 billion links. Cf. e.g. Kevin Kelly: www.kk.org, and Ray Kurzweil's explorations of exponential change: http://www.kurzweilai.net/the-law-of-accelerating-returns.

22 This refers to the potential of a general IT-organized logistics.

available data stock of all humankind. To put it plainly: while in 2002 humankind possessed 2 exabytes of explicit data (i.e. all libraries combined, all newspapers, all music recordings, all videos at all levels), that number had gone, by 2009, to an impressive 250 exabytes. Besides that, in the single year of 2009, more data were generated than all of humankind had ever pronounced (60 exabytes).[23] This gigantic mass of data is indexed by Google. That index supplies, within a fraction of a second, to every point in the world, a list of documents sorted to any combination of search items. Keeping in mind this achievement of Google's, one realizes that the—in many ways not always unjustified—complaints about the quality of search results are nevertheless off the mark. When we imagine that the indexing algorithms recursively become themselves part of any interrogation and search, we may get an idea of the scope of the substrate on which we are able to evolve intellectually, quite differently from today's analytical world. Yet now, in IT and its applications, there emerges externalization of the intellect.

But let us now turn to the going technical articulation of reterritorialization. We start with the sequence 1 0 0 1 1 0 1 0, whyever and wherever generated. One of the above-mentioned processors picks it up. Processors are complete Turing machines, and therefore capable of manipulating the sequence in any imaginable logic, i.e. they are capable of construing any logic apparatus. Generally the apparatus of the processors are built (or, actually, the processors are coded) so as to be able to send as well as receive. The receiving part we described above. On emitting, a bit sequence goes to an analogue-digital converter,[24] which results in the switching-on of certain properties on specific electric contacts of a semiconductor, or a piece of electronics. The processor, with its electrical inputs and outputs, and an actualized logic, has now become an electrical apparatus. So-called actuators are then connected to the outputs. The description is symmetric to the above-mentioned sensors, and may therefore be brief. Optic, acoustic, thermic, kinetic, or comparable actuators convert electric quantities into physical effects. These effects now are no

---

23   This calculation is based on a suggestive train of thought: supposing a person speaks so and so many hours a day, and there ever lived till today, according to today's knowledge, so and so many people, and the daily spoken output were digitized and multiplied by the total living days of the total-ever world population, the outcome would be the said 60 exabytes.

24   DA converters, however, are rarely used, as the skill in electronics consists in keeping bit sequences recognizable so as not to lose them in the entropic analogue noise of energetic movements. Bit sequences may quite easily be intentionally smoothed by so-called capacitive or inductive loads. Something similar happens when electrical devices convert electric pulses to different phenomena. Their mass is sufficient for smoothing impulses, e.g., of thermic or kinetic phenomena. The senses of living beings, too, "smoothen" pulsed physical phenomena. Light-emitting diodes, for example, are actuated at high frequencies, beyond our cognitive range. In order to reduce brightness, pulsing frequency is simply being reduced.

longer in a proportional relationship as was typical for machines, nor in a transformatory relationship as typical for apparatus, but in a medial relationship typical for applications.²⁵

⤦ [FIG. 52] P. 47

DOUBLE ARTICULATION Before presenting examples of printed articulations, let us put our reflections into the perspective of the present discussions about IT. What seems important to remark is the fact that in most such discourses differences are being cultivated with a view to establishing—with or against the epistemic proportional order—a diastemic order that is transformable at will. Above, following our sun language game, we spoke of projections into the reflections of the paths of the sun shadows. But precisely in this discursive cultivating of an arbitrarily transformable order, the emphasized reflection turns, as referential plane, projective itself. In this projected projectivity, we must learn to cultivate differentials, so as to find a way out of distribution, towards a choreostemic order, with, or against, the diastemic order of the large centralized apparatus. If we were to abstain, in the actual energetic (hereafter to be demonstrated in more detail) and logistic wealth, everything would be balanced against everything. That would mean a global entropic super-machine. The sun shadows would projectively be completely illuminated. This is indeed the tendency to be observed in the security, control, design, and expert systems. We also observe medialized projections from the secured realms of shade: mass consumption, information overload, climate catastrophe, proposed philosophic frames of reference, e.g. Wallerstein's *World-Systems Analysis,* Hardt and Negri's *Multitude: War and Democracy in the Age of Empire,* Derrida's "différance"—be they melancholic, agitative, supportive, exigent, whatever. By contrast, choreostemics requires thinking in and with differentials, in their operative double-articulation of de- and reterritorialization dynamics, just as diastemics wants thinking in and with differences, and epistemics, thinking in and with identities and their proportionalities.

IV   TECHNICAL ARTICULATIONS

THE IMAGE For a start, let us take a simple application that clearly shows the way familiar views are being differentialized by new technical articulations. A painting has an identity. A photograph, a printed image in a magazine, are expressions of projections,

---

25   Computers that are so generally configured as here described, are called "embedded systems". Mobile phones, nav devices, washing-machine controls, fuel injections, anti-blocking systems, digital clocks, or blood-pressure meters are embedded systems. Currently, an "electronic Lego" known under the name of "Wiring" enjoys an impressive success among non-IT folks. It helps even non-experts to build rapidly devices that convert physics into code, and code into physics.

of optical differences between image and object, of mechanical differences between print and image. On a flatscreen monitor, however, it is the differential that is constitutive. There are two planes: the concrete, visual plane of the printed colour pixels that may mean anything but are just pixels, and the plane of the represented shapes, texts, pictures, which do mean something but strictly speaking are nothing. Between the two planes—differing from the analytic, continuous, and the logically linguistic times—no relationship may be assumed in an unambiguous fashion on which one might depend. Monitor screens with their pixels as well as the image in its quality as differential are being double-articulated.[26]

↖ [FIG. 40] P. 44

But how may digital images gain stability, how may they be trusted, if no longer thanks to the shadows, the reflections of analytics, the familiar functions, figures, elements, categories? Well, through repetitions, through habits, through exercise. One pixel would not be thinkable, printable, marketable, financeable, without all the other pixels on the many appliances, in the various applications. Not without the many other related pixel technologies, not without the widespread research into semiconductor or polymer doping. The pixel plays on habits, habits play on pixels. The images too actualize themselves on the many monitor screens, must constantly be repeated, play out the most diverse economic or content contexts, get loaded with meanings. They form endless series of nested double-articulations that stabilize, in differentiating reiterations, into populations, and may not adequately be viewed singly, in the conceptions of identities or similarities.

TRANSFER   We are not really fond of that openness of differentials. We find it taxing, upsetting. Therefore, physical and electrical phenomena are treated, IT-wise, as if they were analytic apparatus. The fax machine or mobile phones are such applications. Deterritorialization and reterritorialization are being used exclusively for carrying information from territory A to territory B, just like a letter by post, without modification. The fact that, on the way, all manner of transformations are necessary, and many artefacts are produced, is being ignored, or considered a necessary evil. Nevertheless, the applications emancipate themselves as their usage in the new milieu increases, and new habits arise rapidly, for which new infrastructures crop up. Mobile telephony and Internet telephony such as Skype are good examples. Hardly anyone still remembers that an answering machine, a telephone directory or even a tariff counter used to be freestanding appliances not very long ago.

↖ [FIG. 54] P. 47

---

26   Vilém Flusser, in his *Towards a Philosophy of Photography*, provides an adequately abstract and far-reaching introduction to the challenges of technical images. Vilém Flusser. *Für eine Philosophie der Fotographie*. Edition Flusser vol. 3. Andreas Müller-Pohle, ed. Verlag European Photography, Berlin 1983.

How then is 1:1 transfer from point A to point B coded? How is the processor of a scanner, a fax machine, a printer, a digital camera, or a flat screen monitor coded? So here comes the typical code for each of those devices, in the popular Java "Processing" dialect.[27]

```
1  void draw [] {
2       PImage b = loadImage["cloud.jpg"];
3           image[b, 0, 0];
4       for [int x = 0; x < width; x+=10] {
5           for [int y = 0; y < height; y+=10] {
6               dPoint [x, y, 10, x+width, y];
7  }}}

8  void dPoint [int x1, int y1, int d, int x2, int y2] {
9       color c = get [x1,y1];
10      fill [c];
11      ellipse [x2, y2, d, d];
12 }
```

The command draw in line 1 starts the image transporter. loadImage defines the territory of the original. image defines the starting point of the transfer process. In lines 4–6, the image is being read point by point, and column by column, in 10-unit steps, in x direction in line 4, and, nested for each position x, the column in y direction (line 5), and transported (line 6). In line 7, the machine is closed down. For transporting the points, there is a separate machine (lines 8–12). In line 8, it receives the coordinates of the current point. Line 9 reads the image's colour value from line 2 at the current point, which is being stored in line 10. Line 11 draws a coloured point at the corresponding spot in the target territory (in this case immediately to the right of the original [cf. x+width in line 6]). But here any Internet address or any address of a memory card, hard disk, TV set, printer or computer may appear. That's all it takes to transfer something from A to B, e.g. an image, as in the above illustration.

THE LETTER   The American Standard Code for Information Interchange (ASCII) was established in 1968 and prevailed in the coding of the familiar letters. A, for example, has the ASCII code in its first 8-bit version 01000001; B, 01000010; a carriage return, 00001101. Each punch on a keyboard generated, in this first version, an 8-bit sequence; such sequences are then packaged, addressed, and sent to their recipient, across the electronic networks. For the addressee to be able to read it, every letter must be treated, e.g. by rendering it on a computer screen. To that end, there are so-called fonts, such as the well-known

27 Cf. www.processing.org

Times Roman. This font interprets the code of the letter A, for example, into an appropriate, 16 x 16 bit pattern, which now may be read as the familiar letter A. So as to make the letter better legible, one may use a higher-resolution, 16 x 16 x 4 bit pattern. One point may now be represented in various grey values, and the outlines are smoothed (anti-aliased). To increase the size of the letter, the code remains untouched, a larger bit-mask for the letter A and the colouring of different points on the screen do the trick. There are comparable procedures for graphics, or images, or films. And comparable procedures exist for coding and rendering sounds and music as well.

↖ [FIG. 21] P. 39

ROTATION   Here follows an example for actualizing code through the movement of a so-called stepper motor. It includes various electrical circuits and is so wired that each circuit may position it at a certain angle. The motor in the illustration has 4 circuits any may therefore be turned in 90-degree steps. For a clockwise quarter-revolution, for example, switch S3 must be opened, and switch S2 closed. To rotate it anticlockwise, switch S4 must be closed instead of S2. For clockwise rotation, the switches must be closed in the sequence S4, S3, S2, S1, S4, S3 etc. Inversion of the sequence results in anticlockwise rotation. Increasing the execution speed of the sequence results in faster rotation of the motor; decreasing it slows it down. It is easy to imagine that an appropriate bit sequence may control these revolutions, and equally easy to imagine that a specific bit sequence will bring the motor into a specific position, e.g. by moving 120 steps clockwise. Stepper motors are used, for example, in printers, where a first motor activates a platen for moving the paper lengthwise, and a second motor moves a print head or a pen crosswise. Thus, every point on the paper may specifically be accessed and printed on. There are hundreds of different printing systems, which are all equipped with x,y-directional kinematics.

↖ [FIG. 28] P. 40

COLOUR   An inkjet printer is equipped with a matrix of tiny kinetic semiconductors, so-called piezo crystals that, under electric tension, change their shape and are conceived for "spitting" microscopic colour droplets, as tiny dots, onto paper.
It is then up to the so-called printer drivers to position the discontinuous colour dots so that seen from a distance, the natural and familiar continuousness of images or script emerges.
Colours are being coded and mixed according to different models. For printing, for example, the subtractive model CMYK (cyan, magenta, yellow, key [black]) is used, and for monitor screens the additive RGB (red, green, blue) model, in accordance with the colour sensitivity of the receptors of the human retina. A point of full saturation has the code 1111 or 255; of half saturation, 0111 or 127; of no saturation, 0000 or 0.

↖ [FIG. 25] P. 40

↖ [FIG. 23] P. 39

I   A FANTASTIC GENEALOGY OF THE PRINTABLE

[FIG. 53] P. 47

[FIG. 37] P. 43

Correspondingly, a red dot consists of (1111 0000 0000) or (255, 0, 0), i.e. red in full saturation, green and blue unsaturated; a medium-grey point of (0111 0111 0111) or (127,127,127); a white point of (0000 0000 0000); a black point of (1111 1111 1111), etc.

It is clear that with the availability of such printers (at a price of less than a worker hourly wages), which are able to colour paper with 10–20 colour dots per millimetre at will, any—as well as any future—script, any drawing, and any image may be printed in whatever combination. We do not realize that with an offset printing machine, or a Linotype, this would, if at all, be possible only at considerable cost of handiwork. Thus, these modern processes are very successful. Barely a place in a city that is not plastered with digital print products. Street signs, nameplates, advertising, surfaces, imitations, phantasms...

MOTION Instead of a print head, the stepper motors may also drive spindles or other tools. There, actualizing code for representation or rendering is no longer about the precise positioning of points, but about the generation of continuous movement. Now, stepper motors need not be driven sequentially for filling columns and lines, but synchronously for generating lines or curves. If the motors X and Y move at the same speed, i.e. motor X executes a clockwise step precisely when motor Y does the same, a 45-degree line is generated in the direction from the centre towards 13:30 hours on a watch-dial. If motor X takes two clockwise steps whenever motor Y does one anticlockwise, then a 135-degree line is created, pointing towards 16:30 hours. If motor X does two clockwise steps, as motor Y takes one anticlockwise, a 112.5-degree line appears. Circles require a precise, and precisely timed, synchronization of the two motors.

[FIG. 35] P. 42

What used to be done by ruler, compass, and other aids for geometric visual construction, has now turned into the algorithmically orchestrated clocking of motors. Not a kinematic ballet any more. Kinematics are, as in Bézier curves, being algorithmically imitated. They are the products of analytical geometry. Primacy of numbers over geometry. Non-Euclidean geometry was articulated as early as the 19th century, or even before. Actual use of these geometries in the rendering performed by numerically controlled machines in the machining of real materials, i.e. almost direct concrete availability of these geometries, has also enormously boosted architects and designers over the past ten years.

FORM So as to allow free spatial movement, the code is being rendered into radial articulated systems rather than linear axial ones. Such machines are primarily used for welding, painting or assembling. Gramazio & Kohler, for example, use a robot for rendering a pixel-image appearance to a brick wall. A white monitor screen pixel's code,

255 and 11111111, a medium-grey one's 01111111, or a black one's 00000000, may be directly translated into the rotation of a brick: 11111111 for a 90° rotation, 01111111 for 45°, and 00000000 for 0°.

Much of the current zest for bionic systems, be they straightforward bio-mimicry, or figurative analyses of biologic phenomena motivated by it, rests on the simple and inexpensive reproducibility of constructs in non-Euclidean mensurational systematics, or "geometries". Seen from the angle of de- and reterritorialization, there is little difference between the digital photo-print of plants and a bionic architectural construction: at most, the type of analysis may differ a little, and the rendering process into concrete physics may be more or less painstaking. In none of the cases of such constructions known to us was the principle of intuitive apperception abandoned. Hence the frequent interest in nature, the organic, materials, the living... A mimicry of the idea of energetic growth, hijacked into the tools of competitive struggle for attractions.

↖ [FIG. 27] P. 40

↖ [FIG. 34] P. 42

FAÇADES Media façades are a different story: they are not so easily renderable into ideologies of naturalization. They use the same code all right, to control, for example, the movements of a robot. But these façades consist of a matrix of light dots and may therefore hardly be interpreted as imitations of energetic growth. Their matrix's light dots are made up of so-called LEDs, light-emitting diodes, semiconductors doped for emitting photons when put under tension. A light dot is composed of differently doped LEDs, for the three basic colours red, green and blue. Via the data cable, the light dot is being supplied with a RGB value. It is (11111111 00000000 00000000) for intensive red, (00000000 11111111 00000000) for intensive green, and (00000000 00000000 11111111) for intensive blue. Red with 50% brightness has the code (01111111 00000000 00000000), medium grey (01111111 01111111 01111111). Mixed colours are composed of various combinations of red, green and blue. Eight bits per colour, as per our example, yield a total of 256 x 256 x 265 = approx. 16 million different colour combinations. Light dots are laid out, say, in rows of 64. In that case, the row's code gets 64 RGB values: ((11111111 00000000 00000000) (00000000 11111111 00000000) ...) SYNC. It is being transmitted via the data cable, and the processors of the respective light dots each grab their values from the proper position in the passing "data train". The end of the train consists of a sign due to which the captured values are being synchronized and applied to the brightnesses of the light diodes.

A media façade will be composed, say, of 1024 such light-dot rows. Data supply to these rows follows a similar procedure: ((code for row 1) (code for row 2) ... (code for row 1024)) SYNC. The processors of the single rows (so-called splitters), at SYNC, pass these data, as per above, to their respective light dots.

↖ [FIG. 32] P. 41

↖ [FIG. 46] P. 45

A very simple logistic principle by which city surfaces are being doped in terms of energy and IT. Harbingers of the fact that our cities are no longer cultivating the idea of divine order, or of energetic-natural growth, but that of competing for attractivity. Cities no longer reflect—they shine.

[FIG. 19] P. 38

[FIG. 50] P. 46

[FIG. 7] P. 35

PRINTERS   Then there is another interesting form of technical actualization for the same code. A radicalizing extension of the above inkjet printer is paper that is not being printed upon, but that is rather its own printing device. It works with microscopically tiny globules that are white on one hemisphere and black on the other. They correspond to the pigment of familiar printing ink and are placed between two foils that are imprinted lengthwise and crosswise with parallel, conducting but transparent strips. One customary format includes 1024 strips x-wise and 768 strips y-wise. If, for example, x-oriented strip 128 on the bottom side of the paper is being charged positively, and y-oriented strip 64 on the top side negatively, an electromagnetic field is created at the paper coordinate (128, 64), and the globules turn their black half towards the positive top-side charge. Inverting the charges results in turning field (128, 64) globules' white side up. By hovering over all rows and columns with specific tensions, a printed piece of paper is being created, with the particularity that the colour pigments may be switched again at will, and that this piece of paper can imitate any printed and printable paper page. Energy is only being used for rotating the globules; otherwise the printed image is stable.

A paper printer. Or printer paper. Itself a print product. No need for printable-upon paper. Very, very cheap. A one-way product.

[FIG. 17] P. 38

[FIG. 26] P. 40

[FIG. 12] P. 36

MIRRORS   Printed micro-mechanics. They are capable of tilting microscopic mirrors at very high speed and with precision around an axis. Combined by the millions into a field, they form a special mirror. Strong lamplight is being focused, through a rotating RGB filter, onto that mirror. As one micro-mirror tilts, the overall mirror loses its reflectiveness at this spot. It may lose it for red, green or blue, and it may, since it can tilt at very high speed, also lose it only partially for red, green or blue. Each of the millions of mirror fragments can do it. Simultaneously. The mirror is capable of calling up reflectively, from its discontinuities, any image whatsoever. Its continuous, natural reflection is broken down into invisibly small discontinuities and, for each reflection, getting culturly ("kultürlich" in German, combining "kulturell" and "natürlich") recombined into attractions. Who in the land is the fairest of all?

This so-called digital mirroring device is being produced by Texas Instruments and, under the brand name of Digital Light Processing (DLP), finding widespread use in video projectors, on behalf of, by and large, doubtfully attractive PowerPoint presentations. Regrettably.

SOUND   Piezos (as crystals or semiconductors) distort when put under current, or they produce current when deformed. Like a child on a swing, piezos may be set oscillating by electric impulses, if the piezo's resonance frequency or harmonic frequencies are hit. Such frequency is very stable and well reproducible across a considerable temperature range, as it issues directly from the piezo's crystalline structure. These resonant circuits are used in IT as Units of Time. The frequencies are, for quartz watches 32 kHz, for CD ROMs 45 MHz, for gigabit Ethernet up to 161 MHz. The piezo effect was discovered in 1880, but only began to be understood on a microscopic level by 1972.

A piezo may also be used as a microphone, as it deforms under the influence of sound waves, and produces electric current. With a second piezo, this current is broken down into 44,100 segments per second. An analogue-digital converter measures the tension of each of these segments with 16-bit precision. i.e. 65,536 different pressure levels can be measured 44,100 times per second, and coded. That is the mass of code stored on an audio CD, as well as that translated with an analogue-digital converter into electric currents during play, and transmitted via an amplifier to a loudspeaker or, well, to a piezo. 1:1 de- and reterritorialization.

[FIG. 33] P. 41

Now one might think of using a so-called digital sound processor for combining the coded audio information with others. If a piezo were installed, on the bottom of a cello, every recording would be audible distortedly in the eigenfrequencies of the instrument.

But by coding the spectrum of the instrument's resonant frequencies and working it in inverse fashion into the recording, the piezo on the instrument's body would render the recorded sound neutrally. And in the same way, a piano may be made to sound like a flute, a door like a window, a man like a woman, my car like a concert hall. Not only the tone, but the sound as well may be deterritorialized, into no longer audible, discontinuous fragments, and may be reterritorialized into any acoustic context, and recombined into concrete, continuous audibilies, i.e. rendered.

[FIG. 8] P. 35

MATERIALS   Silver is the optimal electric conductor. Copper, cheaper, is only minimally less so. Therefore it is used in the wiring of buildings. Due to its lower weight, aluminium is used for large overhead power cables. One distinguishes superconductive material, and conductive fluids, such as salt water, which however will not be considered here. Crystalline structures whose conductivity depends on electric current are of particular interest.

↖ [FIG. 55] P. 47

↖ [FIG. 41] P. 44

Semiconductors conduct in specific energy bands, due to a quantum-mechanical effect dependent upon the crystalline or molecular configuration of the semiconductor. The semiconductor may, for example, act as an insulator at low current, but as a conductor at negative or higher current. It is possible to contaminate (dope) the regular structure of a semiconductor with foreign atoms. This leads to so-called hole conduction between the energy bands. Thanks to such targeted contamination, the properties of a semiconductor may be considerably extended. As current passes through a hole conduction, a photon, i.e. light of some specific frequency, may be emitted. Light-emitting diodes (LED), or the new flat monitor screens (OLED) are based on that principle, as are the lasers for reading CDs or DVDs. Or they may, with different doping, heat up or deform when current passes through the hole conduction. Or conversely emit, when the material is warmed up or deformed, electrons of their own, thus generating current. The piezo crystals described above work according to this principle. They are used for microphones, loudspeakers, or for micro-mechanical construction of inkjet printer nozzles.

There exists a large library of specifically doped semiconductors. The majority of sensors for light, brightness, images, temperature, motion, chemicals, "artificial noses" etc., is based on specifically doped semiconductors.

↖ [FIG. 29] P. 40

↖ [FIG. 31] P. 41

↖ [FIG. 6] P. 34

ENERGY   Of particular interest, in a reversal of the light-emitting diode principle, is transformation of sunlight photons into electric current. To this end, a silicon crystal (quadruple bond) is doped with boron (triple bond) and phosphorus (quintuple bond) atoms. As a photon hits a P atom, the surplus free P electron gets lifted to a higher electric potential and finds temporary, and due to particular semiconductor configuration, isolated accommodation with a B atom. If via a conductor, e.g. a copper cable, self-insulation of the energy bands is short-circuited between the B-doped and P-doped areas of the silicon lattice, the electrons flow back to "their" P atoms, and electrical current is generated. This is how photovoltaic cells work.

Simply because we specifically contaminated the crystalline structure of a particular material, it turns into a power generator as soon as we expose it to sunlight. This is fantastic—at once so simple and yet beyond our intuitive imagination. Physicists have learned to see this effect by learning how to deal with mathematics beyond analytical geometry and arithmetic, i.e. with numbers that are not intuitively-geometrically representable. It is the algebraic treatment of symbols that enables technical articulations within the indeterminablenesses of quantum physics—and renders materials printable that generate power when exposed to sunlight. This really is novel thinking and novel articulating beneath the sun. In labs, these effects have been known for a long time. But only now is industrial-scale production about to be mastered, whereby a major problem, interestingly, is the purity of materials and processes. But once the process works, these novel materials may be printed like newspapers. This is why the new print products, such as CD ROMs, processors, data storage, flat screen monitors, and, well, light-emitting diodes and photovoltaic cells too, go down in price 30% every year, and double their range of application every other year.

These developments have two characteristics that are unknown in the production of tangibles such as food, handicraft, or machines. For one thing, their price is not dependent upon their complexity or the number of constructive elements, but exclusively upon the size of the print run. Secondly, on the strength of point one, we see exponential, consistently underestimated expansion of the availability of print products.[28] Doubling of availability every two years means that, if today about 0.5% of worldwide energy demand is coverable through solar technologies, in 16 years, i.e. 2026, the total worldwide energy demand, and in 2032 eight times the world's demand can be covered through solar techniques ... provided we want to. Particularly since the sun's power is equivalent to 10,000 times what we need today. Year in, year out.

[FIG. 5] P. 34

So much for printables. So little, what we actually articulate with it. So timorously do we cling to the familiar reflections of sunlight. Such humility within the shadow's borderlines. So much infuriation over narrowness. So little courage to step out of the shadow.

V ALWAYS ON

A simple intellectual experiment. On the day-side of the earth, we install a photovoltaic foil. Over a very long wire, it sends power to the night-side, where we connect a light-foil to it. It shines, because it is day-time on the other side. 12 hours later, it's day-time for the light-foil, which does not shine, since it's night-time on the other side. Another 12 hours later, the foil is on again, as it's day-time at the other end ... and there is always light at this end!

[FIG. 47] P. 45

Just thanks to two foils and a wire lying about somewhere in the world. No more wood-piling and counting winter's days. No more asking when we wish the light to be on, but just when we want it out. Not primarily logics, but logistics. No more sureness from to following the sun.
But then, what? Whence surenesses?
Prior to script, one relied upon materiality, upon the tangible, directly knowable. Things were mythical, speech only as true as the person whom one knew as speaker. In order to establish certainty over distances and time, a stick was being broken in two, symbolically, so as to

---

28 Cf. e.g. Kevin Kelly: www.kk.org, and Ray Kurzweil's explorations of exponential change: http://www.kurzweilai.net/the-law-of-accelerating-returns.

↳ [FIG. 43] P. 44

↳ [FIG. 15] P. 37

recover, through matching the pieces, the old surenesses of the mythical things once one, or one's descendants, met again.

With the advent of phonetic script, the book replaced the sticks. We no longer rely upon things as such, but upon things' speech, and we check them against visibilities, against things' reflections under the sun, through a combination of geometry and logic. Certainties are being obtained through what we may hear and see within a range of, say, 30 metres. Beyond those 30 metres we cannot hear, and must greatly mistrust our seeing: the shimmering of light, the artefacts of lenses. As soon as we quit the familiar medium of the air: refractions, diffractions, distortions, before our very noses. Few surenesses on the far side of narrow boundaries. When one separates, a book is being written about the reflections of things, in order to recover, through reading the book, the old reflections, on the occasion of one's, or one's descendants', next meeting. Everything would be in order. That could be fine.

And today, with electric writing? Now our shouting range is planetary. Within the 0.1 seconds sound takes to travel 30 metres, the electric signal takes in the whole globe. Indeed, we are, all of us, anytime, able to speak to all the others (today, via mobile telephony, to 5 out of 7 billion people anyhow), no matter where we are, and without wires. And in actual fact, some of us can talk to whomsoever is willing to listen. What, today, are our surenesses about whom to trust? We discover that the reflections of the shadows won't do, that they lead to totalitarianisms, that we must grant more attention to the shimmering and to refractions, that the reflections are all right for surenesses within hearing range, but not for technically supported worldwide ones.

Take winegrowing, in the views and rhythms under the sun that we call nature. Every day, we look upon the vineyards on the hills, and up to the weather, and we read from it what is to be done. We possess words and notions about it, and talk about what is the agenda of the following day, or the following year. The wine will turn out great if we are in optimal sync with the rhythms under the sun, with what we call nature, on the reflecting planet.

And all the winegrower may possibly harvest thanks to this nature is at best 0.2% of the solar energy. And only where the ground is fertile. Territorialized.

But the same winegrower has planted photovoltaic cells next to his vineyard. They are of a different nature. Their signification for us does not follow from intuitive rhythms of the sun. They can be measured and weighed. And they throw a shadow. They even function in accordance with the rhythm of the sun. But it's like an open book one can't read. It too may be held in hand, weighed, and has measurable shadow outlines. But the meaning of the book cannot be apprehended in this way. Likewise, the meaning of these printed cells is not discernable on the strength of their weight, surface, colour or smell. An open page of a book we have learned how to weigh, but not yet how to read.

For there is a crucial difference between the printed book and the printed foil. The book speaks about the reflections and continuities of things. For this, we dispose of words and notions, of well-trained certainties of geometry and logic. Foils, however, open up discontinuities, attractions, the shine of things. Few words and notions are at hand for that; we are as yet untrained in dealing with logistics and operationalities.

And all the while, through this culturly system, we are even today capable of harvesting 20% of the energy radiated by the sun upon each square metre of our territories. Even in deserts and on the seas. Deterritorializedly.

↖ [FIG. 4] P. 34

A rarely experienced abundance within reach. On the way to the shining planet.

INDEXICAL MARKINGS OF THE TOPICS DISCUSSED

These summary discussion threads relate to the lecture, "Architectonics of Narrative Infrastructures", which was presented by Ludger Hovestadt in the first Metalithicum Conference and which forms the basis of this text.

A first discussion thread developed around a certain discrepancy in thinking regarding technology. This discrepancy was pointed out by Hovestadt as unfolding between the abundant space of possibility *factually* posed by technology today, and in contrast to that the diffusely constraining "futurist-realism" way in which our current global problems such as water and energy shortage, hunger, population growth are discussed in political discourse. What lacks in such futurist-realism perspectives, went the discussion, is a consideration of the specific potentials of information technology. Related, and directly opposed to this critique, one contribution considered that this IT-abundance in potentials might need to be axiomatically constrained, that it might be necessary to assume that there is a definitive, solution-realistic limit to possibilities that simply ought to be politically "implemented". But it was also discussed how the discrepancy exposed by Hovestadt comes from the limits of habitual thinking, which presumes that technology would, in order to function, *depend* on a natural predictability of developments. The reference to this familiar thinking would explain why the indicated global crises are conceived and handled in general consciousness as something inevitable, like a force of nature—so as not to disturb this habitual trust in technologically supported uniformity and stability. In contrast, Hovestadt's presentation presents the perspective that predictability was never "natural", but has always, in its conditions, been symbol-technically established and econo-politically administered. The discrepancy in thinking about technology, according to his view, unfolds from the inadmissibility of the assumption of a natural reference system for evaluating the possibilities posed by technology. Generally speaking, the discussions focused on how to identify an engineering-philosophical perspective from the deadlock of a utopian-technistic image of technology on the one hand, and a merely negative-apocalyptic image of technology on the other.

A second discussion thread concerned the concept of a pre-specific abundance of potentiality, as Hovestadt has introduced it, as a political stratagem. As one that comes in from the side of engineering, so it was discussed, it seems nevertheless able to intervene an unusual way in the various political discussions. Other political stratagems were distinguished and discussed comparatively, such as that of an axiomatically conceived idea of limiting, directed against rampant speculation and the risk of hubris, or that of a concrete utopia, in its power-analytic and revolution-dynamic aspects. In many such political stratagems, technology and its possibility space plays a central or even, as a frame set, a *referential* role. Usually this possibility space is administrated by the engineers as experts of what is considered feasible in a realistic manner. It was discussed that in contrast to such a management perspective, Hovestadt's suggestion essentially represents an engineering-philosophical one. The notion of abundance in potentiality does not reference a surplus or a deficiency which needs to be "managed", but rather relates to a space of potential that is provided by information technology infrastructure as a pre-specific space, and that can be "staged" and "populated" in many ways. The discussion revolved around how such a concept of "wealth-in-potentials" could be described in greater detail, from what it can be deduced and how it can be qualitatively specified and distinguished. Especially, it was of interest how to turn away, when considering the idea of abundance, from the concept of an "unproductive expenditure" suggested by Georges Bataille, and his philosophy of excess, to conceive instead something like an extraordinarily productive—a meta-productive—expenditure. In various suggestions, it was considered how this concept of abundance could modulate the traditional language games of creativity, work, and duplicating or re-production.

A third line of discussion developed that revolved around the necessity of an acculturation of this abundance in potentiality. At issue was the question of how such potentiality-laden conditions of daily life may be communicated into familiar ideas. As a central point, there was a discussion of the difficulties that arise from the cultural habit that language is commonly used as—to quote from the discussion—"a window into the world of things being talked about." This habit makes many assumptions that are not adequate for dealing with imaging technology and the use of models as illustrative simulations. The big and rather general question, as has been highlighted, is how an adequate imaginative-analytical thinking could be characterized and developed, one that can deal with the manifold ways of "speaking-by-numbers".

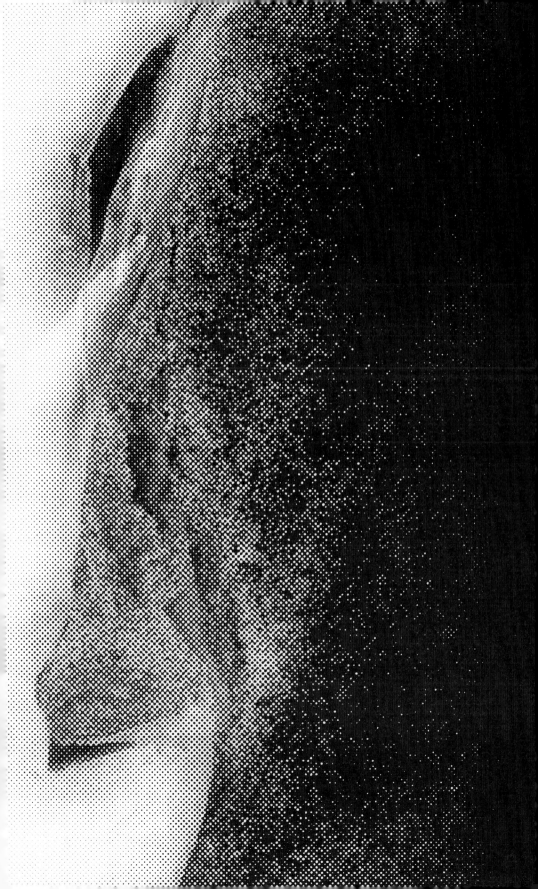

II TECHNOLOGY AND MODALITY [↗ P.70]
HANS POSER

I MODALITIES AND THEIR IMPORTANCE 74 — II TECHNOLOGY AND NECESSITY 79 — III TECHNOLOGY AND POSSIBILITY 87 · TECHNOLOGICAL POSSIBILITY—EPISTEMIC OR ONTOLOGICAL? 88 · ACTUALIZABILITY 89 · ELEMENTARY AND THEORETICAL TECHNOLOGICAL POSSIBILITY 94 · THE POTENTIALITY OF AN ARTEFACT 97 — IV HOW TO DEAL WITH CONTINGENCY 98 — V EPISTEMIC-TECHNOLOGICAL POSSIBILITY 99 — VI FICTION, REALITY AND ACTUALITY 100 · VIRTUALITY, REALITY AND ACTUALITY 101 · VIRTUALITY AND POSSIBILITY 104 · THINKING IN NEW MODALITIES 107 — VII FIRST RESULTS 110

HANS POSER completed his Staatsexamen in mathematics and physics, and his PhD and Habilitation in philosophy. From 1972 until his retirement in 2005, Poser was professor for philosophy at the Berlin University of Technology. Among his main research areas are the philosophy of mathematics, the philosophy of science and technology, as well as recent and modern history of philosophy. His main publications include: Zur Theorie der Modalbegriffe bei G.W. Leibniz. Wiesbaden, Steiner 1969. Wissenschaftstheorie. Eine philosophische Einführung. Stuttgart, Reclam 2001; René Descartes. Stuttgart, Reclam 2003. Gottfried Wilhelm Leibniz. Hamburg, Junius 2005.

Modal concepts as *possibility, necessity*, contingency and reality belong to the most important means of reflection. They constitute philosophical systems—but they are not used in a systematic way to characterize technology. The central ontological problem consists in the fact that technology is based on new ideas, which at the beginning are a mere *possibility*, because the intended artefacts and processes never existed up to that moment. Even the blueprint expresses a *possibility*. But these possibilities must be *realizable*, since technological artefacts or processes have to work properly in the world following physical and causal *necessity*. Moreover, *feasibility* (a kind of *conditioned possibili*ty) and virtuality (as a media reality) have to take into account conditions of the real world (as material, energy, local conditions), cognitive conditions (theoretical knowledge, know how—i.e. *dispositions*, which are possibilities, too), social and cultural conditions (norm, values, i.e. *deontic possibilities*). They all constitute the realm of *technological possibility*. Within this region the development of technology takes place. But at the same time they have influenced our thinking and our culture from the very beginning.

I MODALITIES AND THEIR IMPORTANCE

In a short passage of his essay on Form and Technology Ernst Cassirer develops his view of technological modalities in a very comprehensive way. Technology, he states, "obeys nature and its laws and regards it as steadfast prerequisites of its causing; but irrespective of this obedience against the laws of nature, nature itself is in the view of technology never something ready, a bare set, but rather something

[1] This paper goes back to a fruitful stay as a visiting professor at Rice University, Houston, TX. I want to express my gratitude for this possibility and for many fruitful discussions.

permanently *newly to be set up*, and again and again to be configured. The mind measures the objects always anew against itself and itself against the objects. [...] The further this movement grows [...], all the more it feels itself as mastering the reality. This internal growth does not simply result under the permanent direction, dictation and custody of the actual; but it demands that we return permanently from the 'actual' into an empire of the 'possible' and perceive the actual itself under the picture of the possible. The winning of this view and perspective means, in purely theoretical regard, perhaps the largest and most memorable achievement of technology. Situated itself in the centre of the area of the necessary and persisting in the view of the necessary, it discovers a vicinity of free possibilities. These express no vagueness at all, no merely subjective uncertainty, but rather they step up to the thinking as somewhat thoroughly objective. Technology does not ask in its first place about that, what is, but rather after that, what *can* be. Nevertheless this 'can' itself designates no bare acceptance or conjecture, but rather it expresses an assertoric statement and an assertoric certainty [...]. In this spirit an in itself existing fact of the region of the possible is in a way transplanted to the actual."[2]

As this passage clearly shows, modal concepts belong to the most important concepts of philosophy because they are the base of each complex philosophical argumentation, even if the early Carnap, as well as Quine and Goodman always attacked them. In particular, each philosophical system depends on them, in that it distinguishes between the necessary, the actual, the merely possible and the impossible—and this regarding the levels of logic, ontology, epistemology, of the physical world and of the area of norms.[3] *Logical modalities* concern propositions, where possibility is defined as being free of contradiction, which presupposes a formal language. *Ontological modalities* concern a state of affairs, which presupposes a universe of discourse or a world (up to a world of ideas) including lawlike relations between the objects. *Epistemic modalities* express an attitude of the subject to a proposition, which demands a kind of knowledge. *Deontic modalities* express the evaluation of something, especially of an action, which needs values or norms. Concerning modal statements, Kant highlights that they "contribute nothing to the contents of the judgment, but rather indicate only the value of the copula in relation to the thinking

---

2   Ernst Cassirer, "Form und Technik", in: Leo Kestenberg (ed.), *Kunst und Technik. Wegweiser*, Berlin 1930, pp. 15–61. Reprinted in Cassirer, *Symbol, Technik, Sprache*. Meiner, Hamburg 1985, here p. 81.
3   To speak of the "actual" instead of, for example, the "real" indicates that—thinking of technology—the intended view is always connected with the idea of an action, more than of a res (reality). As a consequence, I will speak of "actualizability" instead of the common "realizability".

in general."[4] All this has as a consequence that modalities are of interest even today.[5]

It is highly important to remember that in most cases there is no way to define modal concepts by non-modal ones (otherwise there would have been no reason for Kant to take them as categories).[6] Naturally, there are logical connections between them within each level; but at least one of them has to be taken as an undefined fundamental category. Modal concepts, therefore, have to be seen as transcendental conditions of our reflection—namely concerning the relation of the subject to its judgments, as Kant puts it. Let us therefore have a look at the logical structure, followed by some remarks on the specific levels which Kant has used, in order to understand their application to technology. The traditional Aristotelian logical connections within one level are as follows:

1 $Na \equiv \neg P \neg a$
   $Ca \equiv Pa \wedge \neg Na$
   $Ia \equiv N \neg a$
   N = necessary; P = possible; C = contingent, I = impossible

This means that N_l, N_o, N_e, and N_{ph} (i.e. the logical, the ontological, the epistemological and the physical necessity) differ in their content, but all of them owe the same structure.

Between them holds furthermore the so-called *modal decline*:

2 $Na \rightarrow Aa \rightarrow Pa$
   A = actual[7]

These logical equivalences and implications are used up to the Wolffian School in 18th century. —Possibility, therefore, is always given if something is actually given and naturally if it is necessarily the case. But in many situations, the term is meant as *mere possibility*, expressing that a state of affairs is neither necessary nor existent, but can exist; in

---

4 Immanuel Kant, 1797, *Kritik der reinen Vernunft*. KdrV B 184.
5 Since Saul Kripke has developed a semantics of modal languages, the reproach that this is senseless has gone. Much more, a broad literature has been the consequence—but it deals neither with causality and physical necessity nor with modalities of technology. As an example for the discussion going on today see e.g. Christopher Peacocke: "Metaphysical Necessity: Understanding, Truth and Epistemology", *Mind* 106 (1997), pp. 521–574, discussed by Peter M. Sullivan: "The 'modal extension principle': a question about Peacocke's approach to modality—response to Christopher Peacocke", *Mind* 107 (1998), pp. 653–660. Peacocke's answer: "The Principle-Based Conception of Modality: Sullivan's Question Addressed", *Mind* 107 (1998), pp. 847–849.
6 Today there are at leas three kinds of position—the most radical one eliminates modalities as senseless, the moderate one seeks for a way of reduction, whereas the last one accepts them as describable so to say from the inside. It is this last view which I intend to address for my purposes. See Theodore Sider; "Reductive Theories of Modality", in: M.J. Loux and D.W. Zimmerman (eds), *The Oxford Handbook of Metaphysics*. OUP, Oxford 2003, pp. 180–208.
7 Here—as in Aristotelian logics—the modality of actuality is connected with an existential import; this is indicated by writing "Aa" instead of "a".

the traditional logics, this has sometimes been called *a false contigent*. As we will see, it is this mere possibility, which dominates concerning technology as long as it has the state of projecting and planning.

In contrast to the Aristotelian connections, modalities of the Megarian type are linked as follows:

3    $Na = Aa = Pa$,

which means: what exists, exists necessarily and cannot be different, i.e. it is the only possible one. Or to put it differently: if something is really possible, then it exists necessarily.

Concerning the deontic modality, scheme [1] is used by Aristotle as well as, for example, by Leibniz, but it deviates from [2] as well as from [3]. If some action a is morally obligatory and therefore deontologically necessary—$N_d\ a$—then one can in no way be sure that it really happens—the only connection which holds is as follows:

4    $N_d\ a \to P_d\ a$
     N_d = obligatory; P_d = permitted; $_a$ = action.

In Kant's *Critique of Pure Reason*, a reinterpretation of modal concepts takes place, which had devastating consequences in the German tradition, for Kant explicitly says that necessity, actuality and possibility are modalities of such a kind that none of them can be defined by the others—not even by both of them. However, Kant in fact makes use of several modal levels, which cannot be reduced to one another—but within those levels, as I have shown elsewhere, he uses the classical relations of modal logic:

absolute or inner modalities:	$N_a\ a = P_a\ a = A\ a$
Megarian modalities	
logical modalities:	$N_l\ a = \neg P_l\ \neg a$
Aristotelian modalities	
synthetic-a priori modalities:	$N_s\ a = \neg P_s\ \neg a$
Aristotelian modalities	
hypothetical modalities:	$N_h\ a = \neg P_h\ \neg a$
Aristotelian modalities	

In fact, these different modal levels are not definable by each other, since they are connected with entirely different ways of reflection: the divine level of the absolute, the logical level of the logical consistency, the categorical-a priori level of the synthetic a priori, and in the end the physical level of the hypothetical, causally contingent connections. Their correlation within Kant's system is guaranteed as follows: the logical level delivers nothing but the structure, which the others have in common, whereas concerning the content the categorical-a priori level is the fundamental one, i.e. its epistemic modalities are the presupposition of the ontological modalities.

The absolute modalities, on the contrary, are dealing exclusively with the absolute, namely with God and the region of ideas.

The distinction of logical, epistemic (de dicto) and ontological (de re) modalities and the separation from absolute modalities on the one side, from deontic modalities on the other shows how extended the spectrum is; furthermore, hypothetical modalities can appear at the same time on all these levels.

Another preliminary remark is needed, because frequently modalities are by no means as obvious as in the cases of the just outlined so-called *pure* modalities; nearly all English words ending -able, such as "breakable" or "computable" express a possibility. In general, all dispositional concepts are precisely hypothetical possibilities—and this at places where we are inclined to speak of sturdy "material constants", such as the melting point of a metal or the rupture strength of a material. This happens in such a natural way as if it concerned nothing but simple statements of fact, because the modal side is intimately connected with a content component. Possibility is also hidden behind each "can", as Cassirer denotes, whereas a "need" as well as a "must" indicates a necessity, often enough implicitly.

A final remark: the basic modality is actuality. There is clearly no way to define it; the discussions on "Being and existence" from Parmenides up to Heidegger show this difficulty. Indeed, we have to take it as *factum brutum*, as a manifest giveness and as a firm destination of existence—not only in a conceptual way. Nevertheless, it is conceptually imbedded in all the other modalities, which combined span an intellectual net of systematic borderlines. So, actuality is to be taken as an occurrence; thinking of technology includes artefacts, their processes, networks and whole systems, which incorporate a conceptual as well as an ontological side. This shows that actuality is not fixed to objects in the traditional ontological sense here: while a water supply system might consist of connected objects and is characterized by a clear process, it does not hold for the software of a computer, not to speak of the Internet as an open system. Another difficulty that cannot be discussed here consists in the fact that—in creating new growing and living entities—bio-technology needs a completely different ontological approach as known from Aristotle and Leibniz. To say it with Ladrière: "The difference between a natural object and a technical object is thus made clear by the difference between *genesis* and *poiesis*. They are two modalities of *provenance*, that is to say of that process by which a thing comes into existence".[8]

All this should help in developing a sensibility for the different ways of how to express modalities regarding technology, and for a better understanding of the complex modal connections to be observed there. However, since today's technology is interwoven with each element of human life, it

---

8   Jean Ladrière, "The Technical Universe in Ontological Perspective", *Philosophy and Technology* 4.1 (1998), pp. 66–91, p. 75.

is not surprising that modal concepts—and in the first place the concept of possibility—are not only used and connected with each other in most different ways, but are accepted so naturally that up to now there have barely been attempts to look at them more carefully, although technology is sometimes called the "art of the possible".[9]

These short remarks give a hint on how to deal with technological modalities as a very specific way of reflecting on technology as an essential part of our lifeworld, as Husserl calls it. Not the formation of a kind of metaphysics of technology is meant here; the intention is rather to have a better understanding of what is going on in the use of modal terms in technology—in order to understand this use as a fundament of our modal thinking—and its development as a means to see and to transform the world. To begin with, we have to look at modalities intimately connected with technology in a quasi-phenomenological way in order to bring them together and to distinguish their different modal levels. This is done in a phenomenological way in so far as we start from modalities in connection with technology as part of the lifeworld, followed by an analysis of conditions which belong to consciousness. To do so, we have to analyse their immanent structure on each level to tease out how the different levels were linked in our thinking. But the guiding idea will be how these modalities— as a means of reflection and as transcendental conditions we impose on it—constitute in a very specific way our understanding of the world.

At present, an entire treatment even of the pure modalities would be taking it too far; therefore, first of all I try to gain an overview concerning the different forms and levels of technology dependent modalities. Even if this centres around the concept of possibility, I will start the quasi-phenomenological view from the concepts of necessity in technology. One cannot expect a system of technological modalities—what is intended, is a clarification of the use of modal concepts in technology and its influence on the development of modal thinking as a way in dealing with our actual world.

II TECHNOLOGY AND NECESSITY

Technical problem solutions must be possible—otherwise, they are good for nothing; anyhow, they are by no means necessary in the strict sense, because other problem solutions are always possible as well. Naturally, since mathematics and logic are used in technology as in every

---

[9] Christoph Hubig, *Die Kunst des Möglichen*, vol. I: *Technikphilosophie als Reflexion der Medialität*. Transcript, Bielefeld 2006.—Exceptions are (along with Hubig) Ernst Cassirer and Hans Freyer; however, their intentions are completely different ones. See Hans Freyer: "Über das Dominantwerden technischer Kategorien in der Lebenswelt der industriellen Gesellschaft", in: *Abhandlungen der geistes- u. sozialwiss. Klasse der Akademie der Wissenschaften u. der Literatur in Mainz* (1960). Reprinted in H. Freyer, *Gedanken zur Industriegesellschaft*, Hase u. Köhler, Mainz 1970. In his analysis of the ontology of technology, possibility is discussed by Ladrière, "The Technical Universe".

science, these a priori necessities—normally the object of philosophical analysis—have to be presupposed; yet there is no reason to discuss them here. But there are technological necessities, which Heidegger summarizes as follows: "Wherever ends are pursued and means are employed, wherever instrumentality reigns, there reigns causality."[10] These causes are of a very different kind:

1. Ontological conditions of *physical necessity* ("physical" here meant as "physis", depending on laws of nature) concern technical artefacts and processes immediately, namely as a precondition: beyond causality technology is inconceivable—which does not exclude considering stochastic processes for technical use, except under special conditions. Technology "obeys nature and its laws and regards them as steadfast prerequisites", as Cassirer has underlined. This physical or causal necessity is not as trivial as it seems to be, thinking of the countless attempts to construct and to build a perpetuum mobile—even after Descartes had formulated his principle of preservation of motion and Leibniz had introduced the principle of conservation of energy. Altogether, physical necessity is taken ontologically—it holds, even if we do not know the laws behind it, since—strictly speaking—we only possess hypotheses. Physical necessity does not simply concern states of affairs—even here, they are taken as contingent facts; therefore Kant and others understood physical necessity as a hypothetical one: it expresses that a state b will causally occur *if* a state a is given. To understand causal necessity as unconditional, means attributing it to natural laws and by this to the processes, which they describe as unavoidably happening as long as nothing interferes. Up to now, there has been no sufficient formal description of this kind of implication, since it has to include time (the present) not only as a parameter, but as the difference between the unchangeable past and the open future—an element that has no place in physics, whereas it is highly important for technology and its orientation towards future.

Even if for these reasons physical laws are not seen as a priori ones today, they have to be understood from the technological side at least as a necessity, which cannot be overcome: if a process in nature is completely disordered (i.e. neither chaotic in the sense of deterministic chaos nor in the sense of a stochastic process following statistical laws) it would never be a candidate for a technology, since technology depends on means which lead to a clear-cut end.

---

10 Martin Heidegger, "Die Frage nach der Technik" (1954); "The Question Concerning Technology", in: Robert C. Scharff and Val Dusek (eds), *Philosophy of Technology. The Technological Condition. An Anthology.* Blackwell, Malden MA / Oxford 2003, pp. 252–264; here p. 253.

So, the question here is not whether the universe obeys causal laws, as Leibniz and Newton believed, or whether a Kantian transcendental subject imposes the category of causality to the phenomena—it is the practical a priori condition, as Arthur Pap tried to understand causality. But concerning technology this has to be taken in an even more restricted way: only in so far as causality is given, is technology possible. In this sense this physical necessity as an ontological necessity is in a quasi-Kantian perspective the condition of possibility of technology.

2 Nevertheless, causal necessity alone is not enough: an artefact must function properly, otherwise we would not take it as the machine in question, but as scrap metal. Because even if a machine produces scrap, it works causally—but not for its intended purposes. What one demands is therefore a *purpose fulfillment necessity* $N_{pf}\, a$. We must exclude, as we used to say, disturbances (i.e. contingent events), so that the causal mechanisms "fulfill their task". Therefore the connection between an artefact and its purpose is an unconditioned, absolute or categorical necessity—namely of a purely conceptual nature, because a technical artefact is essentially defined by its purpose. (One might object that a car that will not start still remains a car; this may be—but if we take its essence to be a means of transport, then this is clearly not given at the moment. Essence, therefore, is not meant in the scholastic ontological way, but what we ascribe as artefact.) The peculiar difficulty lies here in the fact that this conceptual coupling of the artefact or process in question with its aim suggests that there is only one way to reach the intended purpose: namely the way which is incorporated in the present artefact as a means; however, what is meant here is the much narrower relation between the specific artefact and its specific purpose linked in each case with it—and not a global means-ends relationship. (If a special Cadillac is nothing but a means to demonstrate how rich the owner is, there is no need to be a means of transport; so the essence is the way in which we attribute a means-ends relationship to the artefact.)

Furthermore, there is a global aspect, too: a *function* is seen in technology as a means-ends relationship. The function of a type of object or procedure as a means consists in bringing about a type transformation, being seen as the goal. Now, each element of an artefact or of a process has necessarily to fulfill its function. We can speak in this case of a *function fulfillment necessity* $N_{ff}\, a$. The function fulfillment necessity, therefore, holds for the whole (as a type) as well as for its parts (as types)—and will be the guiding principle on the way of development from planning via conceiving and projecting to the elaboration. Concerning the prototype as the last step of elaboration, special means (and not types) have to be attached to abstract functions—means that warrant the purpose fulfillment of the corresponding sub-goals in a technologically effective way. Therefore both these kinds of necessity belong to the absolutely central and essential modalities of technology.

Now there is an interesting point concerning this purpose fulfilling modality: necessarily each existing artefact or process a as well as all its single elements or parts a_i have to have this fulfillment property, otherwise we would not call this the technology in question. So it holds not only that $N_{pf}\, a \to Aa$, but also $Aa \to N_{pf}a$, and therefore $N_{pf}a = Aa$. Let us ask what holds for the corresponding possibility, and let us concentrate on the question of whether we can think of artefacts or processes that do not exist, but are possible in the sense of having the purpose fulfillment property. The outcome must be that we have to negate this. The situation is just that of Kant's 100 possible taler: one cannot buy anything with them; they do not fulfill the purpose of money. A "possible fulfillment property" only makes sense if it is real: speaking of a property fulfillment of something has an existential import, otherwise it is senseless. Naturally—all artefacts are planned as a possibility in such way that *if* actualized, the property fulfillment is given; but this is not the point, here. So, putting things together, the outcome for this modality is that it is of a Megarian type! But in differing from the Megarian ontology, one does not believe that real existence, necessity and possibility collapse—the whole discussion only dealt with the sense we have to attribute to artefacts and processes as their essence. But essences—namely in this case the purpose fulfillment defining the artefact—belong to semantics. Therefore, the modality in question not only differs formally from the Aristotelian ones as it is a Megarian one. It also differs completely from all the others in its universe of discourse: here, the universe of discourse consists of such concepts, which connect specific objects taken as means with specific states of affairs understood as ends under the heading of a specific purpose: therefore this kind of modality is only de dicto, being a semantic one.

3 General living and survival conditions of all of humankind are necessary conditions which are to be fulfilled only by means of technology—as already Plato pointed out in his *Protagoras*: for humans as deficient beings technology is a *necessity of life*. This is a *conditioned* necessity, because technology is necessary *if* we want to preserve human life on earth. It presupposes that—for moral reasons—we are obliged to make sure that humankind survives—which Hans Jonas led to his *Principle of responsibility* as the formulation of his new, future-oriented categorical imperative. But

the necessity of life leaves open which kind of technology we need and which we should avoid. Moreover, we meet a completely different position, if one follows Ortega y Gasset, who explicitly takes his way from "life—the necessity of necessities" as part of a deep-rooted analysis of this kind of necessity.[11] And: "Technology is not man's effort to satisfy his natural necessities. [...] Technology is a reform of nature",[12] namely as an adaptation of the environment to the individual; but strongly speaking, most of the technological objects and processes are "obviously unnecessary" concerning an "organic necessity": technology is superfluous,[13] since "human necessity" is something totally different, namely depending on man's "desire to live well"; much more, "wellbeing is the fundamental necessity of man, the necessity of necessities"—and this "concept of human necessity is fundamental for the understanding of technology",[14] which at the same time creates culture, thus something contingent: the conditionality expressed in both modal theses (and made explicit by Ortega)[15] becomes evident. Anyway, the reasons for this or that interpretation shows that the necessity in question is a deontic modality, and its universe of discourse consists in actions.

4   A new type of necessity is the so-called *factual constraint* (Sachzwang)—an inherent necessity, given by the state of affairs, often used as a persuasive strategy, for example to justify political actions. But even in computer science, an interesting PhD dissertation by Simon White from Aberdeen tries to find a formal characterization of constraints.[16] In his *Abstract* he describes his first goal as to "demonstrate the use of constraint technology to *specify* necessary conditions for achieving a problem-solving goal." Here, constraints are seen primarily in a technological context, whereas it has already been taken in a wider context by Helmut Schelsky years ago. He spoke in a much broader sense of "material compulsory legality" (Sachgesetzlichkeit). With this, he meant the social compulsions originating from new technical solutions people have brought about, which then cause new, unexpected problems, and ought to be mastered only by means

---

11   José Ortega y Gasset, *Meditatión de la técnica* (1939), trans. as "Thoughts on Technology" (1961), reprinted in Carl Mitchem and Robert Mackey (eds), *Philosophy and Technology. Readings in Philosophy of Technology*, Free Press, London 1972, pp. 290–313; p. 291, repeated p. 293.
12   Ibid., p. 292.
13   "Technology is the production of superfluities", ibid., p. 294.
14   Ibid., p. 293 and 294.
15   Ibid., p. 291.
16   Simon White, *Enhancing Knowledge Acquisition with Constraint Technology*. PhD thesis, Aberdeen 2000. http://www.csd.abdn.ac.uk/publications/theses/downloads/white/white.pdf.

of further new technologies.[17] This kind of necessity is explicitly set in parallel with "physical compulsion" (*Naturzwang*), namely as a "self-relation of the person" enclosing all its social and spiritual areas of life. In any case—a factual constraint depends on a view induced by technology, taken with respect to action requirements; that is, a *deontic action necessity* is diagnosed with reference to a technical solution. However, from moral obligation only follows a request to act; and to derive a specific material compulsion from it, seen or used as something to resemble causal laws of nature, and therefore is ideology. In any case, the social compulsion requires a technological problem solution—but of what kind this procedure might be (by the way, again modal concepts: "compulsion", "require", "might"), is not fixed, but open; therefore, the ideology based on so-called factual constrains tries to insinuate a quasi-causal necessity for the proposed solution, whereas it has to be treated as a deontic, not as much as an ontological modality.

Under the heading of this point one can integrate the often quoted self-defence of Robert Oppenheimer, who declared in regards to the development of the atomic bomb, "the reason that we did this job is because it was an *organic necessity*".[18] Davis Baird interpreted it as "two central features of the autonomy of technology. In the first place it recognizes a kind of autonomy. There is a necessity here. But it is not a logical necessity or an *a priori* necessity. It is an organic necessity. I understand this to mean it changes over time and it changes in response to our decisions about our technologies."[19] What is meant is just the factual constraint, which does not need to be understood as a causal necessity, but as a kind of quasi-organic development, depending on the given circumstances.

A comparable kind of necessity is discussed under headings as *technological imperative* in two or three different forms. Best known is what is ascribed to Frederick Taylor stating "Do what you can" or "Is implies ought".[20] This would belong to deontic necessity if there were not such good arguments against it, because each moral is constituted by rules saying what it is forbidden to do just because we can do it—from the Dekalog on. Already long

17 Helmut Schelsky, *Der Mensch in der wissenschaftlichen Zivilisation*. Westdeutscher Verlag, Cologne / Opladen 1961; p. 16f. Reprinted in: H. Schelsky, *Auf der Suche nach Wirklichkeit*. Diederichs, Düsseldorf / Cologne 1965.
18 Speech to the Association of Los Alamos Scientists, Los Alamos, NM, 2 November 1945—Italics H.P., http://www.honors.umd.edu/HONR269J/archive/Oppenheimer-Speech.html
19 Davis Baird, "Organic necessity: Thinking about thinking about technology", *Techné* 5.1 (2000), http://scholar.lib.vt.edu/ejournals/SPT/v5n1/baird.html
20 Frederick Taylor, *Principles of Scientific Management*. Harper Bros., New York 1911.

ago, Lewis Mumford argued against this immoral imperative;[21] but concerning biotechnology, the discussion began anew, namely in concerns of the question of whether we should accept the technological imperative in medical cases: there we have to follow it— that is to say to do what we can do to help in the sense of the old dictum from medical ethics: *Neminem nocere, bonum facere*. Just this argument shows that only those biotechnological means that obey this rule are meant, i.e. that obey deontic necessity. But this converts the whole implication to "Ought demands is".

An extreme kind of technological imperative is meant by *technological determinism*. It says either: "We always do what we can" as a description of facts, or it draws the consequence, pointing out that there is an inner dynamics of technology steered by its own inner technological necessity—and not by humans and their intentions. It had been Jacques Ellul, saying that modern technology does not seek for a means to an end, but has become an end in itself. Technology, he says under the heading of "Autonomy of Technology," presents itself as an intrinsic necessity, "[t]echnique's own internal necessities are determinative. Technique has become a reality in itself, self-sufficient, with its own laws and degenerations."[22] Three points are important here: first, Ellul speaks of *laws*—namely laws concerning the intrinsic necessity of technology, describing the autonomy as a self-organization.[23] Second, they determine as a consequence the "sociological reality".[24] Third, this includes new moral rules the individuals have to obey.[25]

21 He is always quoted as follows: "Western society has accepted as unquestionable a technological imperative that is quite as arbitrary as the most primitive taboo: not merely the duty to foster invention and constantly to create technological novelties, but equally the duty to surrender to these novelties unconditionally, just because they are offered, without respect to their human consequences." Lewis Mumford, *The Myth of the Machine*, vol. II: *The Pentagon of Powers*. Harcourt Brace Jovanovich, New York 1970, p. 185f.

22 Jacques Ellul, *La Technique ou l'enjeu du siècle*. Armand Colin, Paris 1954. Engl. *The Technological Society*. Vintage, New York 1964, p. 134, and Jacques Ellul, *Le système technicien*. Calmann-Lévy, Paris 1977. Engl. *The technological System*. Continuum, New York 1980, passim; abridged selection in: Robert C. Scharff and Val Dusek (eds), *Philosophy of Technology. The Technological Condition. An Anthology*. Blackwell, Malden MA, Oxford 2003.—For a short overview see Daniel Chandler, *Technological or Media Determinism*, ch. "The Technological Imperative", 04/11/2000. http://www.aber.ac.uk/media/Documents/tecdet/tdet07.html.

23 "This autonomy will get its institutional face in self-organization." Ellul (1980), quoted from Scharff, p. 392.

24 "I believe that there is a collective sociological reality, which is independent of the individual. As I see it, individual decisions are always made within the framework of this sociological reality, itself preexistent and more or less determinative. I have simply endeavored to describe technique as a sociological reality." Ellul (1964), Author's Foreword to the Revised American Edition, p. xxviii.

25 "Technology demands a certain number of virtues from man (precision, exactness, seriousness, a realistic attitude, and over everything else, the virtue of work) and, a certain outlook on life (modesty, devotion, cooperation). Technology permits very clear value judgments (what is serious and what is not, what is effective, efficient, useful etc.). This ethics is built up on these concrete givens". Ellul (1980), p. 338.

Examples of today are modern technological systems such as the Internet: as mathematical models of such structures show, the complexity of these systems excludes predictions as well as steering possibilities. So, they are following their own kind of technological determinism as a new kind of non-physical causality. It is not the place here to discuss this position—but if it were meant in a radical sense, there would be no space for freedom and creativity, both of which are essential anthropological conditions of technology: therefore this kind of determinism has to be seen as a conditioned one, depending on the complexity of the socio-technological systems. According to Ellul freedom of individuals is always warranted; but actions including political activities are only successful as long as they are in accordance with the laws of technology development in question.

All these technological modalities from the factual constraint via organic necessity to technological determinism are mixed modalities: they address states of affairs and connect them with deontic principles, as this holds for actions. It might be appropriate to introduce for them a new type, namely *action modalities*.

Comparing these kinds of necessity with each other, one meets four different levels of reflection. The first one concerns the causal basis of laws of nature, the second the semantic-conceptual level, the third the telos level of the purposes, and the last one the level of action in a social context—reaching from deontic modalities, which presuppose free will, up to kinds of necessity which determine individual as well as social actions in a new and conditioned form.

What is important here is the fact that in regards to highly different modal ways we usually think of technology without asking for the conditions and presuppositions behind: the different types of modality seem to be so self-evident for our dealing with technology that a specifying reflection appears as superfluous: nearly all philosophers of technology even did not see it. The reason for this seeming blindness is definitely that we are deeply acquainted with these forms of thinking from everyday acting in the sense of an Aristotelian poiesis:

Acting has to obey physical necessity, even if we are not acquainted with its laws; it is enough to have appropriate rules—as in technology. In each case our actions follow the path of purpose fulfillment—otherwise acting would be senseless. Therefore we project this onto each artefact as a means to an end.

Each of our actions always connects ontological and deontological elements, because they combine means and ends.

Necessities of life, including those depending on culture, are taken as an obligation and as a motivation to act in accordance with these necessities. This view is transferred to technology up to the extreme idea that in a given situation there is only one—and necessary—way of acting.

There is a further way to find a systematic connection between technological modalities. Taking all these necessities as limiting conditions for technological possibility, we arrive at the following conclusions: the widest condition is *logical necessity* as conditio sine qua non and as constitutive for the mathematical or more universally speaking, for the formal side of technology. There is no need to ask whether this would presuppose a logicistic or intuitionistic foundation of mathematics, since all formal tasks in technology up to the extended use of computers are finite. The narrower condition, then, consists in *physical necessity*. Within technology, it is entirely unimportant whether we possess already complete theories as nanotechnology of today shows. It is enough to have effective rules concerning effects in such a way that they can be used as functions for a means-ends relationship, demanded by the *property fulfillment necessity*. A third and more constricting condition of technological possibility is given by the *factual constraints*, which respect the material and social limitations. If these conditions can be interpreted as laws of the socio-technological system, as Ellul sees it, these laws, describing the autonomous self-organization of the system, have the state of *practical and deontic necessities*. Summing up: Each further analysis of technological possibility concerning new and actualizable ideas has to respect this concentric framework.

Taking these correspondences as a systematic background not only for technological modalities, but for a general understanding of technology, now the task will be to fill in this frame and to extend the view including technological possibility, virtual reality and contingency as a next step.

III   TECHNOLOGY AND POSSIBILITY

A technology must be possible—otherwise it is good for nothing. From the necessity condition $N_{ph}\ a$ of laws of nature for any technology a follows $P_{ph}\ a$: the state of affairs or the process a is physically possible according to [1], since it does not offend these laws—but such kind of possibility may satisfy a physicist who accepts Einstein's thought experiment of an elevator in the completely emptied universe; yet, for an elevator manufacturer it is senseless. Technological possibility has to be understood as something much narrower. Let us approach it gradually. The problem Ernst Cassirer points at, with reference to Friedrich Dessauer, is the most difficult modal difficulty of technology: before an artefact is real, it is nothing but an *idea*, thus nothing but a *purely imagined possibility*, which owes no other kind of actuality. So, it is so far a *mere possibility*. And in fact, speaking of a possibility in a technological context, it is always a mere possibility, not yet actualized and with no necessity at all. Therefore, the whole discussion that follows presupposes this narrower concept of possibility.

However, differing from artefacts of fine arts and from fictions in literature one expects that this possibility in technology can be brought to actuality, namely as an artefact or as a process. "Can be brought to actuality"; this again is a modal marking which is precisely *actualizability*—a modal possibility P_t which needs further analysis. Naturally it would contradict [2] to believe that $P_t a$ would imply a, but it is meant as something which can be brought to actuality by human actions. There are trials to develop a formal system for these connections, using a kind of action-operator $B\,per\,A$ and elements of tense logics.[26] This indicates that actualizability is a very strange possibility. It must have interesting properties, since all our technical acting (such as planning and expecting the end) depends on it. The human being is not only as Benjamin Franklin meant, a *tool-making animal*, but a *creative being* producing with regard to technology new ideas, which are actualizable as absolutely new things and processes in a spatiotemporal actuality.

TECHNOLOGICAL POSSIBILITY—EPISTEMIC OR ONTOLOGICAL?

The short remarks show that the concept of technological possibility demands two clarifications: What is meant by this possibility? And: How to understand actuality? The two concepts are interwoven—but it makes sense to start with the clarification of this kind of possibility in question, which will be enriched step by step by a discussion of actualizability. Possibility can be understood in two different ways, which partly mark the difference between the Megarian and the Aristotelian modalities. If one understands possibility of a state of affairs a as its complete set of conditions (so that no quality and no relation are left out), the conclusion drawn by the Megarian Diodoros Kronos has been—as already mentioned—that the existence of a is necessary. There are two ways out. One consists in Leibnizian possible worlds, which are built up by complete concepts of all the individuals belonging to one world: among them God can choose the best one; and freedom of human individuals including their actions and their development of technology is warranted, since this is part of each complete concept of each individual. The other solution goes back to Aristotle. It takes possibilities as a set of conditions, which are not complete, but open: a trunk can be used as a beam or as firewood or, or ... All these possibilities are inherent possibilities of the trunk.

Now what about a technological possibility? The Megarian way, as well as the Leibnizian, have to be excluded, since our problem actually consists in the fact that we develop different possibilities—sometimes completely new ones—and then decide which to actualize. What kind of possibility is this? It cannot consist in a complete set of

---

26 Klaus Kornwachs, "Zur Logik technischer Entscheidungen", in: Hans Poser (ed.), *Herausforderung Technik*. Peter Lang, Frankfurt a.M. 2008, pp. 131–160.

conditions—this would exclude the possibility of a selection between them. If these possibilities are taken as epistemic ones we would have to discuss propositions; if they are ontological ones, possible states of affairs (artefacts, processes and systems) are meant. Ladrière speaks of creative "information", given as a "project representation" and later on transferred into an "operating representation".[27] This is what he meant by inventions, but it holds for each foregoing idea, which has to be fixed in a representation: the possibility then has reached the actuality of a plan up to a blueprint. Nevertheless, the ontological status remains a hybrid one, namely an epistemic possibility as information, and an ontological possibility thinking of the intended artificial object.[28] Since a reflection on technological possibilities aims for a choice between mere possibilities, which to actualize, they must be taken as ontological ones; but they are given only by symbols—namely as propositions or as a plan describing these possible artefacts, processes and systems—symbols that presuppose the necessary knowledge of how to conceptualize (not yet to actualize) the possibility in question. These descriptions can never meet their object as individual ones, even if they are sometimes meant that way (e.g. a tunnel under given unique geological conditions). Therefore these possibilities are concerning *types*, not tokens. Otherwise it would not be possible to teach and to learn technology, since what will be taught and learned are rules concerning functions that connect ends with means. Thus, technological possibility differs from classical ontological and epistemological possibility, because it bridges the two and includes always categories of means and ends, which belong to neither one side nor the other.

ACTUALIZABILITY   Let us now turn to the problem of actualizability. In philosophy it has already been taken as a huge challenge already throughout centuries. It already arose when Plato sought the relation of ideas to phenomena; as is well known, he introduced the demiurge as a craftsman, who had to accomplish the transaction of ideas in spatiotemporal actuality. Aristotle secularizes all this, so to speak, and discusses it as *poiesis*, bringing-forth. In the Christian tradition the question had been how to understand the divine *creatio ex nihilo*. In modern

27   Ladrière, "The Technical Universe", p. 76.
28   I reject Ladrière's reductive thesis that "what is called 'thought' is only another name for this [inventions producing] cerebral activity. It is, to be sure, indisputable that the activity of thinking is possible only in connection with cerebral activity." (L.c., p. 77). If we understand thinking as depending on a complex neuronal structure (used today as one of the models), we might be able to explain why new ideas can occur; but there is no way to explain within these models how and why we are free in choosing between possibilities in a rational way—but free will belongs to the anthropological presuppositions not only in technology, while complex systems cannot be steered to strange attractors whatever they might be known or not.

times it took the form: how to bridge the two ontological areas divorced by Descartes as *res cogitans* and *res extensa* in general, and in particular between a totally new idea and its material factual actualization as a thing? Thinking of Leibniz and likewise of Christian Wolff (the inventor of the term "technologia"), this transition is accepted as a *complementum possibilitatis*, namely as the divine "Fiat!" as a power of the divine mind, which has to be added to the purely conceptual regio idearum. But Kant dismissed this solution with mockery. Nevertheless, the four Kantian concepts of possibility cannot help here, because he takes possibility with respect to the spatiotemporal world as actuality minus concrete situations (the possible is that which is real at some time) or as an imagination (*Vorstellung*);[29] if one would start from there and refer to the fact that the technological idea will be real at some time, even if this time lies in the future, one would have to assume—completely against Kant's intention—that everything, including what we see as a new and creative object, is already given in the region of ideas—a Platonian and Leibnizian position, which Dessauer accepted: for him, the engineer is a discoverer, who finds the solution design in a Platonic world.[30] The metaphysical price of such an eternal Patent Office is not only very high—Dessauer's proposal does not help; for what would be the criteria for separating working solutions from illusory ones, science fiction from something actually actualizable? Above all—what should be a human "fiat" procedure? Of course, to look at technology is a look on the part of possibility, because we intend an alteration of the sphere of spatiotemporal actuality: the engineer is not a "realist", as he himself might believe, but as the Austrian poet Robert Musil states, a possibility person, someone thinking of possibilities. Just this very specific possibility concept assumes no Platonic world of ideas, no Leibnizian region of truths of reason, or even of possible worlds. However, it requires a deep-rooted clarification including the element of time in particular.

Getting closer to an understanding, it is advisable to change the viewpoint and not to look from possibility to its actualization, but from actuality back to the prior possibility. This is so to speak a technology-oriented view of possible worlds in the manner of Jaakko Hintikka, who wondered how to reach a possible world from the existing one by means of concepts.[31]

29   "Das Schema der Möglichkeit ist die Zusammenstimmung der Synthesis verschiedener Vorstellungen mit den Bedingungen der Zeit überhaupt [...], also die Bestimmung der Vorstellung eines Dinges zu irgend einer Zeit." Immanuel Kant, 1797, *Kritik der reinen Vernunft. KdrV*, B 184.
30   Friedrich Dessauer, *Streit um die Technik* [extended 4th ed. of *Philosophie der Technik*, 1927]. Knecht, Frankfurt a.M. 1956.
31   See, for example, his "The Modes of Modality", *Acta Philosophica Fennica* 16, pp. 65–82. Reprinted in Hintikka, *Models for modalities: Selected essays*. Reidel, Dordrecht 1969. Also in: Michael J. Loux (ed.), *The Possible and the Actual. Readings in the metaphysics of modality*. Cornell Univ. Press, Ithaca NY 1978, pp. 65–79, here p. 67.

Let us start from the *already actualized artefact or process*—as both existing and functioning in the sense of purpose fulfillment necessity; of course the causal necessity is assumed and fulfilled. Heidegger speaks of "Her-vor-bringen", bringing-forth, explaining: "It is of utmost importance that we think bringing-forth in its full scope", and he adds that this bringing-forth is located in the craftsman: that is it what he calls "Entbergen", revealing, which encloses "the possibility of all productive manufacturing".[32]

Chronologically, looking back from the actuality to the prior possibility we meet the blueprint of the yet not existing artefact, but fixed in all its details (forget at the moment that even the blueprint is an artefact—we take it here as a complex and complete symbolization of the not yet existing artefact; the same holds, if it is not drawn on paper but on the screen of a computer). Each element is registered without already being a machine—it concerns an actualization possibility, which will guarantee at the same time the purpose fulfillment necessity, when the symbolized content of the blueprint will be actualized. But how do we know this? There are three levels of answers:

— The *easy answer* is a classical school-oriented technological solution: the artefact will function as purpose-fulfilling because we know all components from earlier actualizations, i.e. we know their construction, thus (1) we know all parts, and we know that they have worked in a purpose fulfilling way, i.e. effectively; (2) we know furthermore that all these parts are technologically compatible, in the strict sense that the functioning of the whole is guaranteed, since the existing artefacts consisting of these parts fulfill their function.

By this we come across two modalities, one connected with *function fulfillment* and the other with *compatibility*. The concept of a *technological function* belongs to the absolutely central ones, since each technological explanation and each technological development depends on it.[33] To fulfill a function is in no way solely founded on observation of actuality because besides the descriptive element (concerning what one can observe) it is finality that is assumed, namely the possibility of reaching a given or intended goal or a purpose by the

---

32 Heidegger, *The Question*, Scharff / Dusek, pp. 254 and 255.
33 Up to now, there have been only a rare number of explicit analyses of the concept of technological functions. Günter Ropohl makes extended use of it (see his *Eine Systemtheorie der Technik. Zur Grundlegung der Allgemeinen Technologie*. Hanser, Munich 1979, rev. ed. 1999), but he takes it only "in the descriptive meaning of the word" (G. Ropohl: "Philosophy of socio-technical systems", in: Evandro Agazzi and Hans Lenk (eds), *Advances in the Philosophy of Technology*. Newark, 1999, pp. 317–330. Also as: http://scholar.lib.vt.edu/ejournals/SPT/spt.html, Society for Philosophy and Technology, vol.4.3 (1999), p. 63.

given means.³⁴ As a means-ends relationship, a function allows different means to reach the goal. The other direction holds that a means may be used for different ends, so that it is connected with different functions. Functions are often treated as purely descriptive in biology—but as a function denominates a means-ends relationship, to speak of a function presupposes an interpretation, which understands something given as a means to an end, whereas neither means as such nor ends are observable entities. In biology and medicine the interpretation is done by a technology projection: when the heart is said to have the function of circulating the blood, it is seen as a pump. (Purely causal connections cannot be functions in nature— no one would say that gravity has the function of making the ripe apple fall from the tree. This holds, since the concepts of means and ends have no place in physical processes, because Aristotelian teleology had to abdicate. This does not exclude the use of physical processes as means for purposes—think of a drop ball, depending on gravity; but the drop ball, not the gravity is a means to an end.) The function fulfillment necessity, introduced above, has already shown the importance of technological functions. Precisely this function fulfillment necessity, accorded to existing means to an end, is presupposed in the "scholarly" case of a technological solution: this necessity-connection of something given, seen as a means to an end, is taken as *effectivity* and handled as a disposition. This is not astonishing if one remembers that dispositions are a shortform of a law of nature, here substituted by an empirically approved *rule* concerning a means-ends relationship. (Rules, by the way, are neither true nor false, they are no propositions at all—they are directives that have to be effective, when followed up in a goal-oriented action.) To know the function concerning the relation between specific means and its ends means knowing the rule that precisely these specific means necessarily fulfill the needed function, if actualized. This is the reason why we, knowing the rules in question, can substitute the blueprint by a computer program, which develops the intended construction map: the rules are fixed in its software.³⁵

Technological *compatibility* or compossibility, the second modal term, is much more than mere logical or physical compossibility (i.e. fitting together without contradiction), because it concerns

---

34 Peter McLaughlin, *What Functions Explain: Functional Explanation and Self-Reproducing Systems* (Cambridge Studies in Philosophy and Biology). Cambridge University Press, Cambridge / New York 2001.
35 See Hans Poser, "Programmare col computer in una prospettiva filosofica", *Discipline filosofiche* I.1 (1991), pp. 175–188.

the interlocking of functions in such a way that the purpose of the whole as the final purpose is warranted—which again presupposes the purpose fulfillment necessity on each level. Formally we need further rules of the type $N_{ff}x \wedge N_{ff}y \rightarrow N_{ff}(x \wedge y)$ for means of the type x and y.[36] This shows that we have to think in hierarchies of means, fulfilling functions that for their part belong to a corresponding hierarchy. These hierarchies depend on the final goal of the artefact, process or even technological system. The compatibility of these means concerns not only each sublevel, but also the interplay in such a way that the means on each level warrant by their functions the function fulfillment up to the final level. Formally, this is also expressed in rules, which have to be effective. As a consequence, the argumentative structure of engineering sciences differs essentially from that of sciences of nature.
So, even in this easy case the modalities we meet are intertwined with finality. "Fiat", here, becomes the rule-dominated purpose-guided action of actualization: the efficient rules warrant the actualizability of the possibility and its successful transformation into actuality.

— The *more complex answer* is as follows: as soon as a single supplementary innovation is necessary for a part, because it is required by the customer or legally (for instance an engine with lower flue gas volume of a certain kind), one can optimize this part (this would be the classical action) or develop an absolutely new part depending on an invention (for instance a new catalyst). Both steps are based on purpose-oriented research, by which one finally reaches a purpose-fulfilling part compatible with the whole, which can be expelled in the blueprint as actualizable. The actualizability is based on previous empiric research. Here an enlarged "Fiat" is based on results of ideas-escorted purpose-oriented research, which first develops all new action rules.

— The answer at the *uppermost level* concerns a completely new technology which is based on the brilliant connection of new elements—elements of which actualizability, as well as their compatibility, is not yet known. Here each single element has to be checked as just said in concerns of a more complex answer. The complementum possibilitatis as the "fiat" has just this step as its pre-condition.

---

36 These are only seemingly logical symbols, because the conjunction designates the physical interaction of two means in such a way that their functions constitute a new complex function, a transition that is symbolized by an arrow. To develop a formal system for rules is not intended here; if one wants to do so, one has to add elements of tense and action logic, since one has to distinguish at least between rules that express a co-functioning of two parts at the same time and others that express a functioning one after the other.

All this shows that viewing technological in the possibility perspective is, as Cassirer rightly states, much more complicated than the writing of a new novel. Finally the technological solution always consists in going back to elements for which actualizability is warranted, and for which one has good reasons to expect compatibility, because the qualities in question are already known at least for similar parts. Yet it still holds that the new creative ideas speed up the technological development: the ontological difference between the possible and the actual continues to exist. Yet the attitude in the region of possibility, in the case of technological ideas, is always interwoven not only with the reflection on actualizability, but also with actions of trials for an actualization and—if required—by alterations. Actions of trial and error, according to Karl R. Popper the central elements of empirical research, are here the crucial bridging element: they have to be incorporated into a theory of technological modalities. Where Kant understood his epistemological categories as conditions of the possibility, not only of the knowledge, but also of the ontology of things, we have to take into account that in technology this is immediately connected with actions, and thus with intentions, goals and the searching for means. Actions go on in space and time, and—concerning the thinking and acting subject—they consist in a synthesis of ideas and spatiotemporal modifications of objects. All this holds even if these activities are substituted by machines—yet the conditions and the interpretation of the results go back to human actions and depend on them.

ELEMENTARY AND THEORETICAL TECHNOLOGICAL POSSIBILITY
To speak of possibility presupposes freedom and free will in order to allow new ideas and a rational decision on how to act. Concerning technology this demands two different kinds of reflection; the theoretical one asks for a foundation of this condition, the practical one seeks the ways of technology assessment, taken in the language of technological possibilities: for feasibility.
What has been said can be seen in a further important perspective: the noteworthy point of the different and quite well-known three levels of technological possibility just presented is that they all depend on categories of finality. The use of technology assigned actions to homo sapiens from the very beginning. He learned what techniques in fact fulfill their function, thus which are suitable as means to an end, a finis. And he learned what *human possibilities of actualizability* are—a position differing from a Popperian falsificationism of hypotheses. So it is not surprising that humans can apparently immediately grasp all these different possibility types just listed, and actualize them in actions— seemingly without previous reflection on possibilities. In fact, all our actions are based on rules concerning goals, means and functions, on forecast, on possibilities and on the expected results of an action. This is

it, why the close acquaintance with and acting through technology is to be understood as humans' way of dealing with and being in contact with actuality. Consequently, based on our action skill, this kind of dealing with possibility and finality is so natural to us that it is not experienced as a special reflective act. Therefore, this type of technical possibility might be called *elementary technical possibility* in order to separate it from other forms to be discussed now.

The view just developed is to be seen as the epistemic possibility counterpart to the semantic purpose fulfillment necessity; however, we need to characterize the levels of possibility in a different way. The "easy answer" is already more complex than these elementary technical possibilities: as a theory-laden answer, it is previously prepared by engineering sciences, and might be called a *theoretical technological possibility*. The theory including its rules warrant without any doubt a kind of theoretically founded actualizability. But, as even millions of never actualized patents show, this is far removed from hard facts. It is indispensable to take these facts into account—and just this is expressed by a very new type of modality, namely *feasibility*. No politician renounces this concept in his rhetoric; and if the title of an engaged article on energy problems is "Necessity and feasibility of an energy turn", it is evident that modal concepts are indispensable for a reflection overcoming our dull actuality.[37] The *mania of feasibility*, inspired by a blind prejudice of progress optimism, will be ignored here. Instead, we will turn to *feasibility studies*, today called *project studies*, which are now en vogue.[38] They are really highly important, because they have to work out whether a theoretical technological possibility of some abstract actualizability is indeed actualizable in a given concrete situation and its circumstances. Feasibility, therefore, expresses what we really *can* do; therefore it is sometimes called a *real possibility*.

Just this is the place of technology assessment as a systematic, methodologically guided analysis and evaluation of these possibilities in order to decide rationally what possibilities should be actualized and under what conditions. In this case, the place of the abstract actuality as such is clearly substituted by a very concrete description of the given situation. This includes—despite all complexity reduction— the available knowledge, the actual skill (know-how) of engineers, as well as of workers, the availability of manpower, energy, material and infrastructure, the local geographic, geological and climatic conditions as elements in order to evaluate the possible results and to tax the estimated costs as well as subsequent costs—elements which clearly do not belong to the purely technological actualizability. Thus, beyond the technological actualizability we are confronted with a

---

37 See, for example, Hans Lenk, *Macht und Machbarkeit der Technik*. Reclam, Stuttgart 1994.
38 In Germany, there is already a standard for how to perform these studies according to current practice: DIN 69905.

whole bundle of heterogeneous modal requirements such as juridical admissibility, availability of resources, organizational practicability, economic feasibility and risk consideration. Therefore, it is in fact this form of reflection that brings about the connection between the theory of engineering science (which tells us what is actualizable according to technological knowledge), and what is actualizable in the concrete given situation. This even concerns the actualizability of the blueprint—remembering the immense number of such drafts which, although actualizable in a strict sense, were never actualized. At the same time feasibility studies indicate conditions under which an actualization will be possible if one increases knowledge and know-how, settles manpower, guarantees the energy, and material supply, and so forth. But moreover—feasibility includes ecological, social and cultural conditions, as well as a limitation of feasibility or these conditions demand solutions with reference to factual constraints, as introduced above. Solutions that also have a technological nature (such as the residential and care possibilities accompanying the settlement of new manpower: medical care, the construction of streets, markets, schools, churches, mosques or temples—which again requires further construction workers, businessmen, doctors, teachers and priests). Seen from a feasibility perspective, two borderline cases appear. Some of the suggested technologies were—caused by the given limiting conditions—denoted as *impossible* (e.g. a perpetuum mobile, a time trip for people into the past or the future, the construction of a bridge in a protected area of nature), while, on the other hand, a *necessity* will be claimed to take this or that way as the only passable one. (This, by the way, is nearly a Megarian modality, since the agreed necessity depends on one sole possibility; but as it depends on further conditions, it is the interesting case of a conditioned Megarian modality.). Nevertheless, even if feasibility has an actual representation of the information on how to assign actual material elements to the informational data, it remains a possibility in its characteristically hybrid structure as an epistemico-ontological modality.

All this shows that the modality of feasibility reaches from necessity via possibility down to impossibility. Its new element is the systematic reflection on actualizability in an extended and enriched space of possibilities. By this, the horizon of the human thinking of possibility is extremely widened compared with an unconsidered state of elementary technical possibilities or a restricted reflection. Even if this modality is at the same level as the necessity assertion of the factual constraint, it operates without an ideological touch, since it systematically includes conditions, which reach from the factual given state of affairs and its possible modifications up to the area of the norms and values: in thinking a common measure of them has to be found in order to combine technological and deontic modalities.

THE POTENTIALITY OF AN ARTEFACT   Each artefact is actualized to fulfill its function. But even this is a kind of possibility of the actual artefact, namely having the potentiality or disposition as a means to transform an object or a state of a special kind a into a different one of the kind b as the goal. This procedure, this process is intended, its fulfillment is essential for the artefact—therefore its possibility is, as we have seen, taken as a necessary one. This intrinsic ontological modality can be seen as the transformation of the information fixed in a rule of how to connect the specific means with the specific goal and inbuilt as a finality into the artefact. Or it might be seen as the causal process determined by the steering mechanism as its antecedent conditions. Seen as a potentiality, i.e. as a modality, these two views are brought together. In this way, we understand a technological artefact as an "objective possibility"[39] in order to transform something by a process depending on inbuilt information to an intended goal. Therefore, even artefacts have to be seen in the light of modalities as Ladrière aptly describes it: "thereby, the operation receives a properly ontological import, because the possible, as modality, belongs to Being."[40] Now, this kind of Being caused conceptual difficulties: "But if there is an objective possible, belonging, as possible, to reality itself, this means that there is, in Being itself, a defect of being." The Being, as he sees it, "is held in check by a condition which limits it from inside"; yet I think there is no need to put things this way—even more, one has to accept that we are not thinking in categories of Being, but in potentialities and in dispositional predicates, which we take as if they were manifest qualities. In this way we plan and realize actions, and this is how we form artefacts in order to carry an intended potentiality. The dissimilarity between mere possibility and potentiality consists in the understanding of the latter as connected with an inner tendency towards transformation of the given to a future state. This clearly holds for a machine already running by an energy supply—but it holds even for a stone, picked up as a tool to crack a nut: we then attribute this nut cracking potentiality to the stone, even if the energy supply depends on our hands. Leibniz in his *Monadology* speaks of an *appetitus*, in his works on sciences of a *conatus*, and concerning ontology of an *existiturire* as coming into being of the possibilia. Or as Ladrière puts it: between the genesis of nature and the poiesis of our acting one is confronted with a "parapoiesis" within the "technical universe".[41] This constitutes the remarkable difference, since it involves a dynamic coming timely to being.

39 Ladrière, "The Technical Universe", p. 19.
40 Ibid.,
41 L.c., p. 88f.

IV HOW TO DEAL WITH CONTINGENCY

A special case of possibility is contingency, thus a possibility which is at the same time an un-necessity (or logically equivalent: neither necessary nor impossible)—in colloquial language quite often called "chance".[42] Mostly we are persuaded that—seen in a macroscopic way—chance is an epistemic modality, so that in lacking further knowledge, we are only able to formulate probability statements: We do not know the outcome when shaking the dice cup. But already the case of a decay of a single atom of a radioactive material is an ontological, not only an epistemic case. The same holds for quantum-mechanical processes. Something similar applies to complex causal systems, of which we can already prove in their mathematical model of deterministic chaos, that we are unable to make long time predictions. How, though, shall we handle all that in technology? Mostly contingency is seen as an accident or collateral damage, and by this as undesirable; but more on this later. If above was stated that technological artefacts have to work causally reliable and function-fulfilling, it was meant too narrow though, by reason that there are technologies whose objective is to harness chance. The drawing of the winning lottery numbers shows a machine whose purpose is precisely this; and as a result, like every roulette table, it is regularly checked by specialists to see whether this purpose is fulfilled. Therefore, we have to distinguish at least the following cases:

— In the case of a dime it is about a simple artefact whose purpose is to warrant the same probability for all six of its surfaces when thrown. We try to actualize an "ideal cube" as a dime—and not a manipulated one.
— In the case of a nuclear reactor, as well as with a medical treatment using radioactive materials, we base the technology on half-life, allowing us to harness probability technologically.

Therefore, the presupposition of a physical necessity, made at the beginning, may not be limited to causality in its strict sense—crucial alone is that the lawlikeness warrants the respective purpose fulfillment, so that one can formulate the means-ends relationship as an efficient rule. We all are acquainted with the concept of a "creative chance", which as a *serendipity* is a show-stopper for now. In such cases it is not enough to suppose with Whitehead and many other complexity theorists that nature is also creative, for in technology the outcome has finally to warrant the purpose fulfillment necessity. This results in the problem of something new in a contingent observation taken as a positive assessment. This presupposes that the observation was not intended, but can

---

42 Discussed with respect to technology by Kai Weiss: "Der schöpferische Zufall", in: Alexandra Lewendoski (ed.), *Der Philosoph Hans Poser*. Sand + Soda Publishing, Berlin 2007, pp. 107–109.

be used for a technological purpose. But one can neither wait for the lucky chance, nor locate it, nor teach it.

— However, in concerns of its absolute opposite, the *unhappy chance*, the bad luck situation is completely different. It is taken as the annoying incidence of unforeseen events in a singular case. It is contingent in the same way as the happy chance—but everything is done to avoid it. This shows that *danger* is a modal concept too, namely the possibility that a given position turns over into a position undesirable to us. A real danger is nothing else than an emphatic expression for an undesirable imminent possibility just before its actualization. In technology we go to any lengths to avoid that the purpose fulfillment function might be in danger.

The problem only consists in the fact that the appearance of bad luck is not predictable. Thus, possible *hazardous incidents* are imagined in order to prevent their actualization by suitable technological measures. *Safety installations* go back to the Neolithic period when together with settling down, embankments, fences, ditches and walls were established around the settlements to protect human life. In early-modern mining they can be observed in mining technology, and since the second half of the 19th century they have been systematically developed as safety features and equipment.[43] They reach from the centrifugal auto-advance mechanism via automatic train stops up to concrete covers around a reactor. Thereby a further extension of technological possibility thinking becomes apparent, not only because purpose and function fulfillment have to be part of the blueprint, still equally and in an increasing fashion mechanisms of *contingency avoidance*: not predictable, but possible contingent circumstances ought to be made impossible, that is to be excluded from actualizability. The risk society we are living in is supposedly characterized by just the fact that it is not ready to endure insecurity on a much higher level than any earlier society, just because on the part of modern technology we expect the avoidance from disturbance possibilities.[44]

V EPISTEMIC-TECHNOLOGICAL POSSIBILITY

Both technological and scientific inventions and innovations—briefly, all basic improvements—assume the—not at all trivial—ability of humans thinking something NEW; the new barn gate is an obstacle to

---

[43] See Stefan Poser, *Museum der Gefahren. Die gesellschaftliche Bedeutung der Sicherheitstechnik* (Cottbuser Studien zur Geschichte von Technik, Arbeit und Umwelt, vol. 3). Waxmann, Münster 1998.

[44] Heinrich Lübbe, *Der Lebenssinn der Industriegesellschaft. Über die moralische Verfassung der wissenschaftlich-technischen Zivilisation*. Springer, Berlin / Heidelberg / New York 1990.

the proverbial ox. Humans are creative beings. However, not only that—we must even be enabled to communicate the new to others, maybe by metaphoric extension of meaning in language (*techné* signified in both Greek and Germanic language at first wickerwork, then timber-frame construction, before it came close to us in the extension of meaning). Or it happens by means of totally new concepts, which then need an explanation depending on metaphors, analogies etc. All that requires even for the others the ability to think in possibilities. However, this is—as Cassirer emphasized—probably the basic contribution of technology to culture, because this thinking allows us to look towards the future in actualizable concerns, but not yet existing possibilities including imaginations of new constellations of environment and society.

Just this form of thinking possibility has taken a totally new and enlarged form, namely, as a *thinking of possibility of possibility*. Here Hans Freyer has already alluded to this: "Already the simplest steam engine is [...] no longer a tool. But it signifies only the very beginning of the new line. To provide powers for not yet fixed purposes becomes from that time on the central intention of the technology."[45] In the past, bricks were burned for the construction of this church or that house, often enough respecting the form required in each case — windows of Gothic brick churches are a clear documentation of this. Today, however, parts are produced to be available for any buildings: They are possibilities for possibilities. This becomes visibly even more clear by the computer: the intended purpose is to be open for possibilities—much more, for as yet unknown programs, which for their part are not fixed to a certain purpose, but to *possible purposes*: not a specific, but any workshop drawings can be provided to be printed out as a blueprint (still as a possibility). Today this phenomenon counts for whole production streets equipped with industrial robots which can be fed with new software: they are conceived from the very beginning as an open possibility of a model change. Indeed, there are attempts to iterated modalities already in the Stoic logic—but today's natural use of iterated technical and technological possibilities is a new phenomenon and signifies another substantial extension of the dimension of human thinking and reflection.

VI   **FICTION, REALITY AND ACTUALITY**

The extension of possibility thinking has astonishing parallels in the fictional worlds of cyberspace. Differing from novels of the world literature, they exist in a completely technology-based immaterial region. It is most astonishing that one can decide to have

---

45   Freyer, *Über das Dominantwerden technischer Kategorien* (1970), p. 139.

a double life as Albert Camus postulates by "vivre le plus" or Max Frisch in his novel "Mein Name sei Gantenbein" had imagined—with the serious difference that one can enter this fictional-real PC-life with a new identity and surrounded by a whole community; for actual money one can buy imaginary apartments and equipment, and one can write to imaginary correspondents actual mails that will actually be answered. All this differs basically from theatre or from someone who slips into a play as a "Gothic" in the role of a mediæval person, where helmet, buckler and a knight's armaments are made of actual materials. There has always been such kind of play, even as a play in the play and as a play between joke, irony and deeper meaning. However, the accessories of cyberspace inhabitants are absolutely fictive: technology has allowed Leibnizian possible worlds, existing in the divine region of ideas, to take place in a server or on the computer screens. In a way never obtainable before, picture, imagination and actuality conflate in thinking and feeling. Even if Ray Kurzweil's Nanobots are science fiction—his way of speaking of "real virtual reality" indicates the new way of dealing with these possibilities.[46] And a decade ago Michael H. Heim already developed *The Metaphysics of Virtual Reality*—nearer to Heidegger than to McLuhan.[47] Altogether, everywhere in the world dozens of actual conferences on virtual reality took place. Whole libraries are filled with books on virtual reality, most of them belonging to computer science, many dealing with the relation between information and society. This shows that it is an important theme, which has to be taken into account regarding questions of technology and modality.

VIRTUALITY, REALITY AND ACTUALITY   Problems concerning virtuality begin already with the concept: whereas in German the word "virtual" signifies something completely unreal and purely fictional, in the physics of geomatrical optics it designates a simulacrum; in French one takes it as a (partly dynamic) possibility to act. In English it means the essence of something, though not formally recognized.[48] All this shows that it is a hybrid concept—in particular in connectives as "virtual reality", which for German ears sounds as "frozen heat". Eric Champion quotes from Weckström, who declared: "a virtual world has to support the following factors: there has to be

---

46 Ray Kurzweil, "Twenty-first century bodies" (1999), in: David M. Kaplan, *Readings in the Philosophy of Technology*. Rowman & Littlefield, Lanham 2004, pp. 381–395; p. 387.
47 Michael H. Heim, "Heidegger and McLuhan: The computer as component", in: *The Metaphysics of Virtual Reality*. Oxford University Press, Oxford 1997, pp. 55–72. Reprinted in Robert C. Scharff and Val Dusek (eds), *Philosophy of Technology. The technological condition*. Blackwell, Malden MA 2003, pp. 539–545.
48 "being such in essence or effect though not formally recognized or admitted", Merriam-Websters Dictionary.

a feeling of presence, the environment has to be persistent, it has to support interaction, there has to be a representation of the user and it has to support a feeling of specific worldliness."[49] Heim counts even seven different meanings of "virtual reality", going back to the intellectual fathers—from photo-realism up to fictional worlds, and he concludes: "Behind the development of every major technology lies a vision. [...] The vision captures the essence of technology and calls forth the required cultural energy to propel it forward."[50] In his *Virtual Realism* he uses a comprehensive formulation depending on three characteristics: "Virtual reality is an immersive [i.e. affecting the senses] interactive system based on computable information."[51] He understands this as virtual reality "in the strong sense"—namely depending on "full sensory immersion—no keyboards and monitors".[52] This is nearly a Weberian ideal type; but since the term is used in most instances in the weak sense, speaking of virtuality of reality as well as of possible worlds, I do not restrict the concept to the "strong" sense, here.

Even if "the essence of the American space program [...] comes from 'Star Trek'", as Heim says,[53] it remains the challenge how to link completely fictitious arts with actualizability and actuality. His conclusion is: "VR [virtual reality] promises not a better vacuum cleaner or a more engrossing communications medium or even a friendlier computer interface. It promises the Holy Grail",[54] namely by means of an "esoteric essence" of this technology. Concerning modalities, the interesting point is not so much the esoteric of mysterious dreams as in Richard Wagner's *Parsifal*—Heim's favourite example—but the extension of actuality and its opening to and amalgamation with virtuality. This has always been part of cultures in mystery plays and rites—and it has always been connected with techniques of priests: dances, rites, rhythm and songs leading to a trance which wears away the difference between actuality and virtuality. But today's technology raises new questions, since all the criteria of Weckström are parts of what we use to call actual. Whereas in fact they are measuring virtual reality against the actual, with which we are acquainted in such a self-evident way that no further discussion of it seems to be necessary. So, what is called "real" are elements (and only selected ones) of actuality.

49 Erik Champion, "When Windmills Turn Into Giants: The Conundrum of Virtual Places", *Techné* 10:3 (2007), pp. 1–16; Weckström, N., *Finding "reality" in virtual environments*. Department of Media, Media Culture. Arcada Polytechnic, Helsingfors / Esbo 2003. [could not verify: In other places, Champion quotes it a "thesis" of 2004]
50 "The essence of VR", in: *The Metaphysics of Virtual Reality*, pp. 108–127, p. 117; reprinted in Scharff and Dusek (eds), *Philosophy of Technology*, pp. 546–555, p. 550.
51 Michael Heim, *Virtual Realism*. Oxford University Press, New York 1998, p. 6.
52 L.c., p. 46f.
53 Heim, "The essence of VR", p. 123 / 552.
54 L.c., p. 124 / 553.

Let us keep in mind that it has always been possible taking something for real opposed to our "lifeworld" as Husserl calls it, of our mesocosmic world of actions and experience, of history and of finiteness of life. One might think of Plato's ideas, Plotinus' world soul, Cantor's sets and alephs—but all that does not belong to the modal theoretical phenomena of virtual reality. Naturally—"reality" is a modal term as well as "actuality", which I have been presupposing so far as an unproblematic fundament. Atomic physics and especially quantum theory has challenged this and substituted it by electromagnetic fields, so that the reality of the microcosmos consists of a probability distribution instead of particles, whereas astrophysics is speaking of cosmic worm holes and complete worlds behind it as the reality of the macrocosmos. One has to say what reality is, in comparison with actuality. It makes sense to call my dreams real, to speak of my perceptions differing from the perceived object as real, to say that causality is real, that the legislative or the executive power are real and so forth. Therefore Christoph Hubig made and developed the proposal to take up a classical difference between reality and actuality:[55] Descartes speaks of *realitas formalis sive actualis* as the way in which a res in its nature is given, whereas *realitas objectiva*—differing completely from current use—concerns the res as the object of thinking, namely as an idea. This goes back to the terminology of the scholastics, according to whom we have to distinguish between *actualitas* concerning the res given in the world in its causal acts and connections on the one side, and *realitas* as our imagination of it on the other. Picking up this distinction, it is possible to clarify the concept of virtual reality, in so far as it concerns realitas or reality, but not actualitas or actuality: virtual reality concerns the imagination, which differs from the actual as the involved technology and from our lifeworld. This had been the reason why I spoke of "actuality" instead of "reality" throughout this article: the blueprint as a paper is actual—but its meaning as an actualizable artefact is real.[56]

The reality attributed to technologically generated virtual objects should fulfill Weckström's criteria; this shows how important Kant's forms of intuition are, since as much as possible of them should be fulfilled—whereas this always happens only in part and by a drastic reduction (virtual smells and temperature do not belong—at least up to now—to the area of technological possibilities; the means are concentrated on colours and sounds, presupposing that the user takes

---

55 Christoph Hubig, Lecture *Realität, Virtualität, Wirklichkeit*, University of Stuttgart 2003 / 2004, 12 handouts. I express my gratitude for sending me these materials.
56 An article with the promising title "What's Real About Virtual Reality?" by F. Brokers, one of the leading specialists in this area, describes the technological side as "now really real", but he does not touch the side of virtuality. Frederick P. Brooks, Jr. "What's Real About Virtual Reality?" In: *IEEE Computer Graphics and Applications* 19.6 (1999), pp. 16–27. Online: http://cs.unc.edu/~brooks/WhatsReal.pdf

them as three-dimensional content). This shows that a virtual reality can never be a complete, technologically generated map of actuality in its complexity. Therefore it has the structure of epistemic possibilities in so far as they only represent types, not tokens, and do so presupposing the epistemological categories of knowledge, since without them the user would have no chance to interpret his impressions as dynamic objects.

VIRTUALITY AND POSSIBILITY   Now, technological artefacts and processes belong to actuality. It is trivial that the criteria of Weckström for virtual reality are also fulfilled there, since they stem from the lifeworld and are used to show how small the difference between virtual reality and actuality seems to be. But the difference always remains—imaginations do not act (or only seemingly). Nevertheless, the situation is much more complex, since technologically based virtual realities are of very different kinds:

1. There are ways of planning and developing new machines, ships, complete technological systems, as well as architectural projects for houses and skyscrapers up to whole settlements on the computer.
2. There are computer-based virtual realities which have an enormous impact on our actual world since they are used for education and training—e.g. flight simulators.
3. And there are games that produce completely fictitious imaginary worlds with imaginary actions.

The first type consists in the immediate prolongation of classical engineering: CAD (Computer Aided Design) was originally developed to substitute the work of engineers at a drawing board, so that the whole construction can be done up to the blueprints and part lists by taking into account all technological norms and relevant laws. Now, 3D-CAD produces virtual impressions of the parts, its connections etc., so that the experienced engineer might decide whether the possible problem solution he views on the screen is acceptable or should be modified. All this happens without blueprint, without physical model, without experiments to make sure that compatibility is ensured, because these elements are already presupposed and fixed in the technological rules as a part of the software. Therefore the design can be modified, and by means of DMU (Digital Mock-Up) even functions were controlled and virtual prototypes developed. The forerunner is the automotive industry, where the development time of a new car has now been halved. In all these cases, the technologies depend on an impressive connection between the actual world and a technologically generated fictional world. The computer allows us to develop plans of artefacts, i.e. *actualizable possibilities;* what is new is that all required technological rules, obligatory laws and norms are part of the program; moreover, that it is possible to directly develop alternatives and to make them visible. These last two elements—the active intervention and its visualization—are taken here as virtual reality. The goal of the technology in question is the production of possibilities, namely of actualizable delineations of an intended artefact.

The second type connects virtuality and actuality in a new and specific way: flight simulators are today's standard way to train pilots—and at the same time they are used to test new steering and security equipment before it is installed in actual planes, so the TU Berlin owns such a simulator for just these two purposes.—Quite similar are the conditions of minimal invasive operations in a hospital, where the physician uses electronic steering instruments, first to train it (namely in a virtual, only possible surrounding of imaginated reality) and then to operate actual tools and monitoring his actions on the screen, as if he or she would actually look into the actual body of the patient.—Virtual firms in a virtual society are in use to train businessmen by means of virtual business companies: they learn how to treat a firm, how to react to market movements etc.: A virtual surrounding—visible and seemingly allowing autonomous acting by seemingly actual handles—is an *image* of an actuality (e.g. an airport projection on the wall plus the steering equipment: *both* are seemingly actual, but each one knows that they have the reality of a picture). This image is produced by the actual technology, and the goal is to train specific motor skills (the pilot or the doctor) or to educate someone in how to decide in conformity with practical rules (the banker) in all types of possible situations—namely possible in the lifeworld, so that they all, later on and in an actual situation of the type in question, react in the adequate way. Therefore, possible situations are presented via technology as types, which can be the case actually. These possibilities as virtualities have the function to allow training without risks: the technology that is developed for these purposes owes its actuality the principle of contingency avoidance in order to minimize risks.

The last type, the games in their highly differing forms, produce virtual reality as seemingly possible worlds. There is no need for actualizability, fun is the goal. The fascination connected with these virtual possible worlds is the actual possibility of the player to leave the actual world behind. Taking virtual reality in the strict sense of Heim, the connection to the actual is cut off: "*Virtual worlds do not re-present the primary world. They are not realistic in the sense of photo-realism. Each virtual world is a functional whole that can parallel, not re-present or absorb the primary world we inhabit.*"[57] And this realism in VR, he points out, "results from [...] livability":[58] this modal term designs

---

57 Heim, *Virtual Realism*, p. 47f., repeated p. 138.
58 L.c., p. 48.

a possible lifeworld—Leibniz would have spoken of a possible world with a different Adam—and this allows a reflection on the essentials and the problems of our lifeworld. This "new reality layer brings an ontological shift", he points out, and this is important, here. According to him, it is art which makes use of it: "the art of virtual reality takes us deeper—not into nature, but into technology".[59] My point is that this technology opens new experiences of possibility, even in games and in arts, since it is the way of dealing with possibilities—not only in a conceptual way in a region of ideas, but at the same time in sensory experience and emotions.

All that is done in a fictional way, which means: The imagination is as if the elements were an actual part of a machine or an existing airport or patient or firm, whereas they have in fact only the reality of the mind's eye. The very new element of these virtual realities as possible worlds consists in their way of mapping actuality not only by concepts (as it holds for Leibnizian possible worlds), or by fixed pictures, or moving ones as in science fiction films, but by a dynamic synthesis of forms of thinking and forms of intuition. It is the active and perceiving subject that interprets the data, transformed from their electronic codification into a visual impression, artificial sounds and factual movements (e.g. of the flight simulator), to which the subject reacts actually. All this presupposes not so much theoretical knowledge, but practical experience for the interpretation of these elements as seemingly actual ones, knowing at the same time that all this differs from the history-laden and finite human life. But the *actions* of the pilot in the simulator, of the doctor, the banker as well as of the engineer at their computer—acting just as the player with or without helmet—belong to the actual world. Moreover, these actions have consequences in the actual world: the pilot, the physician and the banker learn their job, namely how to act in the actual world. The actions of the engineer lead to modifications within the technological possibilities offered by the computer, resulting in a new blueprint. And the game player has the possibility of seeing him- or herself as a member of the Star Trek crew. This is just the very new and important element: the results brought about by these technologies, taken as virtual reality, allow an immediate use of possibilities in the actual world. The fascination of chatting depends—in addition to the element of play—on the possibility of earning real money by this or transforming the virtual reality into actuality, e.g. by an actual appointment with the imaginary partner.

Any computer simulation in science and engineering depends on a mathematical mapping of natural, social or technological processes, which produces nothing but model-dependent results, i.e. possibilities.

---

59 L.c., p. 73.

But they are interpreted as if they were actual; consequently they are used for decisions—and in the next step even these decisions might be integrated into the program. Human-machine interface is a harmless name for the connection of possibility, taken as reality, and actuality. But things become much more difficult, if one takes into account the actual influence: all these virtual world elements, i.e. possibilities, seen from the side of actuality, have an effect on the individual psyche and the actual changes in society created by them.[60]

THINKING IN NEW MODALITIES   The question now is how these new elements influence our culture and our understanding of the actual world. The term "virtual reality"—and to attribute reality to its phenomena already indicates the direction—namely to enclose this kind of possibility in actuality. This not only differs from classical modalities—it is a change in our lifeworld; not in the sense of a loss of actuality, but as an extension of the lifeworld. It begins with the question of whether virtuality, distinguished from actuality, can be seen as a possibility at all: to ask for a technological possibility in the above-mentioned sense might be difficult, since the technologically produced results are understood as real ones. Clearly, the technology is actual, it is purpose fulfilling, means and ends are evident. The goal is just to present a possibility as a virtual reality; this shows best in critical remarks, saying that the computers are too slow, the feelings too artificial, not to speak of missing smell... Thinking of the goals, it might be more convenient to speak of an "augmented" actuality, as Heim did.[61] This shows in all the efforts to develop a kind of ethics for actions within virtual reality,[62] whereas moral rules for possibilities might make sense only under the presupposition of actualizing them. At the same time it focuses on a new way to deal with modalities, since this augmentation makes possible worlds part of the actual world! It is the actual technology that allows this kind of augmentation by the means of virtuality. But this differs from Goodman's World making, according to which we live in different worlds simultaneously. These technology-related worlds are "as if" they were actual ones—not in the global sense of Hans Vaihinger's *Philosophie des Als-Ob*, since we are convinced

---

60   Erick Champion describes players of games such as *World of Warcraft* as follows: "They do not buy these games because the games are programmed to have conditions and triggers, they do not play these games because the games are rule-based systems; they play these games because the games *challenge* them to change the world and to explore how these character roles embody and express aspects of their own personality." "Windmills", *Techné* 10.3 (2007), 3.
61   He, in fact, writes "augmented reality"; see his chapter "The essence of VR" in his *Metaphysics of Virtual Reality*, p. 128, reprinted in: Scharff and Dusek (eds), *Philosophy of Technology*, p. 554.
62   A small but concise overview starting from computer ethics up to cyber ethics is given by Philip Brey: "Theorizing the Cultural Quality of New Media", *Techné* 11.1 (2007), 2–18, p. 2.

that there is a difference between virtuality and actuality. However—
what actually is actuality? As long as artefacts had been machines this
seemed to be clear, but what about the processes going on in the hardware, steered by a software? As uninterpreted ones they are meaningless; so they are physically interpreted as an electronic processes, computer technologically as bits and bytes, then, e.g. as coloured pixels, interpreted as a picture of somewhat and finally taken as virtual reality. But this can only be done if we are acquainted with the actual: the virtual reality can be understood as such only if we know what kind of possibility a mere possibility is (i.e. excluding necessity as well as actuality, both of which are trivial possibilities) and how it differs from actuality, to which the technology in question belongs. Technology—and this is the interesting point here—provides means that allow this kind of imagination of something in an actually much more universal way than each traditional literature, picture, theatre performance or film. It is this extension of modal imagination, which, instead of purely thinking in modalities, characterizes this new step into possible worlds as a virtual reality.

Let us focus on the importance of virtuality and its kinds of reality as the amplification of the modality of actuality. The influence of technologies which provide virtual reality has only been touched in rare cases up to now.[63] Even if the problem of virtual reality is not meant in the general sense in which McLuhan suggested already in 1964 that electronic media change our way of thinking and feeling, it is helpful to push his guiding idea forward, according to which "each medium of communication not only carries people or ideas, but also transforms the cultural environment of which it is part", as Edward Relph paraphrases him.[64] Relph adds: "...it is certainly the case that coincident with the recent growth in use of electronic media there has been a huge cultural or post-modern shift that has affected art, literature, philosophy, science, geography, architecture and town planning. In all of these there has been a move away from the objective, rationalist perspective that seeks a uniform account of the world, to a view that acknowledges the validity of many different perspectives." One of these elements, where actuality and reality differ from each other, is the loss of regional bindings. Globalization, pushed forward by the use of the same computer

---

63 As an example, Peschl and Riegler (to whom Champion refers with the accurate statement "our notions of reality are actually cultural notions of a constructed reality" (p. 12)) are interested in the changes in scientific methodology, but touch culture as a whole only in some remarks. Peschl, M.F and A. Riegler, "Virtual Science: Virtuality and Knowledge Acquisition in Science and Cognition", in: A. Riegler, M. Peschl and K. Edlinger (eds), *Virtual Reality: Cognitive Foundations, Technological Issues & Philosophical Implications*. Peter Lang, Frankfurt 2001, pp. 9–32. Online paper at http://www.univie.ac.at/constructivism/people/riegler/papers/peschlriegler01virtual.pdf

64 Edward Relph, "Spirit of Place and Sense of Place in Virtual Realities", *Techné* 10.3 (2007), pp. 17–20, p. 20.

software everywhere in the world, software which for technological reasons is totally independent from the actual place, has destroyed the historicity of places where to build an Italian palazzo—since the virtual world and its places have no such bindings to history, neither of places nor of individuals and societies. Globalization in the sense of getting rid of history widens freedom of choice—visible in the widening of feasibility. But it results in "inauthentic places",[65] where authenticity is understood as the basis of individuality up to societies.

Now it is interesting to observe that this globalization caused by virtual reality, mixed up with or taken as actuality, is balanced in an intense protection of history as world heritage and a growing return to regional traditions. As a reason, Richard Coyne explains: "We don't only think about places, but we think through them. Places seem to function cognitively"; and this holds, since "the environment acts as a source of associations, metaphors and stimuli through which to think".[66] The differences between virtual reality and actuality are obvious: we would lose history and individuality—but since we need both, one might, as Heidegger did, quote from Hölderlin: "But where danger is grows / The saving also" (Wo aber Gefahr ist, Wächst das Rettende auch.)[67] It may be that the use of methods of VR to reconstruct history as Virtual Heritage, argued for by Jeffrey Jacobsen and Lynn Holden, belongs to these antidotes, because its "real goal is to understand ancient cultures".[68] Whether this holds or not is not so important—it is much more important to see the differences between reality and actuality. Heim, in giving a short overview of his book, says: "Realism in virtuality should seek neither photo-realistic illusions nor representations. Realism, in the sense of virtual realism, means a pragmatic functioning in which work and play fashion new kinds of entities. VR transubstantiates but does not imitate life. VR technology is about entering worlds and environments, and worlds arise from the human ability of adapting things through pragmatic functioning." And somewhat later: "The real can no more be reproduced online than it can be replaced by fantasy. Reality is transformed by entering the virtual. Virtual worlds need not suggest a replacement of the primary world, nor should they be so fantastic as to terrify common sense."[69]

---

65 Relph (2007), p. 22.
66 Richard Coyne, "Thinking Through Virtual Reality Place, Non-Place and Situated Cognition", *Techné* 10.3 (2007), 26–38, pp. 29 and 32.
67 Friedrich Hölderlin, "Patmos", in: Sämliche Werke. Grosse Stuttgarter Ausgabe, vol. 2/1, 1951, p. 165; quoted by Marin Heidegger, "Die Frage nach der Technik", in: *Die Technik und die Kehre*. Opuscula 1, Neske, Pfullingen 1962.
68 Jeffrey Jacobsen and Lynn Holden, "Virtual Heritage: Living in the Past", *Techné* 10.3 (2007), 55–61.
69 Michael Heim, "Virtual realism", April 1998, http://www.nettime.org/Lists-Archives/nettime-l-9806/msg00026.html. The complete book: Michael Heim: *Virtual Realism*. Oxford University Press, New York / Oxford 1998.

He is speaking of these new fictional worlds that belong to play and arts, to a world outside work and everyday life. But his remarks are important for the scientific and industrial use of these new worlds, since it is knowledge and know-how, stemming from the actual, which is the fundament of all training programs as well as for all simulations. They allow us to build up an action capacity on the one side, to enroll possible scenarios for a further development of processes, technologies and structures in nature and society as hypothetical possibilities, depending of the choice of the model, its parameters and its starting conditions. Knowing about these differences, technology-based virtual realities are an important enrichment of human thinking and experience. They have widened the horizon of thinking and sensing in possibilities and iterated possibilities of possibilities up to dynamic ones which allow us to reflect on other worlds.

May it be all up in the air if this, as Heim hopes, increases the harmony in our virtual as well as in our lifeworld. But in fact, Hans Jonas' Principle of responsibility demands just this kind of modal reflection on possibilities, since his famous and well-known principle is given in its "classical" formulation as follows: "Act so that the effects of your action are not destructive of the future possibility of such [genuine human] life".[70]

VII FIRST RESULTS

Summing up all these elements, one instantly sees that technology is not a single isolated area of possibilities, but is constituted by the connection of various ones, namely

1. on the basis of logical modalities and semantic modalities as Megarian modalities of purpose fulfillment necessity;
2. we find modalities as such of the physis in its old Greek sense (physical possibility),
3. supplemented by theoretical knowledge (elementary technological possibility),
4. developed further into iterated modalities (possibility of possibility),
5. including factual conditions of the society (as an economic, resources-conditioned feasibility),
6. evaluated by deontic modalities (as a moral, social and legal admissibility including normative ecology etc.), and
7. widened by virtual realities as non-actual elements, interwoven with elements of the actual world.

---

70 Hans Jonas, *The Imperative of Responsibility. In Search of an Ethics for the Technological Age.* University of Chicago Press, Chicago 1984, p. 11.

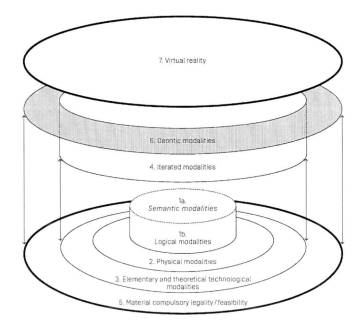

All these areas and levels go simultaneously into technological considerations. At the same time a continual connection is erected between them, for instance, while actualizability acts as bridging modality between epistemic conceivableness and ontological actuality, whereas the modality of material compulsory legality has to hit the bridge between technological possibility, factual actuality and normative possibility qua admissibility or even obligation as a social necessity or possibility. On the level (7) we can observe the intimate connection of possibilities, taken as virtual realities, with actuality in feeling, thinking and acting. Therefore technological categories embrace considerably more than a positivistic understanding of facts, because they must include possibilities and functions, purposes and goals, future perspectives and normative horizons. It is not technology as such which marks the difference between human and animal, but this wide ambit of an orientation by means of modal, creative and self laying objectives in thinking and action.

All this does not solve the philosophical problem of a foundation of actualizability of a possibility. The Platonian solution depends on far reaching metaphysical presuppositions, which in fact cannot solve it. The pragmatic attitude, which seems to be behind the way used here, from the actual back to the possible, would be nothing but a persuasion, but no argument. The cultural historical view, which Cassirer seems to argue for, cannot be the key, since the lock is missing. Putting things together, the situation is as such: in fact, humans are creative, we have enlarged the world by completely new substances, artefacts, processes and network systems, and in fact we can think in possibilities as well as in possibilities of possibilities, and we can do so by bridging different formal structures for different levels of modalities. In fact, modalities are not reducible to non-modal concepts. In order to explain all this, we have to enlarge the presuppositions—as Kant did, when he saw the categories as the unavoidable a priori conditions of knowledge and of ontology, which are constituted by the subject of knowledge. But Kant's scheme is, as we know, too narrow and has no flexibility. Now, Whitehead made a fruitful proposal in speaking of a creative development of schemes of thoughts throughout the history of ideas: our philosophical systems mirror this creativity as well as the invention of mathematical and logical structures, for example in the 20th century. My proposal is to see the foundation in this development towards modal categories as creative inventions of the human mind, which are interwoven with acting, with the thinking in means and ends, with projections to the future and in reflecting on all these elements. This intellectual development has its parallel in acting: the way leads from natural material, used as tools, via the production of tools, the division of labour and so forth up to modern technological systems. All this has allowed the development of these new schemes of ideas—in technology and outside. So, the fundamental presupposition is just the one corresponding to the anthropological preconditions of technology. The deficient being that both Plato and Kapp have in mind, is, to put it in a positive way, not completely fixed on instincts. Therefore humans are free, as Herder explains it, they are characterized by a world openness as Gehlen adds, and they are creative, as Whitehead presupposes for all of nature. Complexity theories of today and their use in describing social systems, as well as structures of our neuronal network, might be taken as a support: the mathematical structure shows that these social and neuronal systems can develop new ordering structures. The mathematical structures—we need to reach these results—are the outcome of formal creativity in concerns of the structures and of creative thinking in terms of technological possibilities concerning the computers and their software, which generated these results.

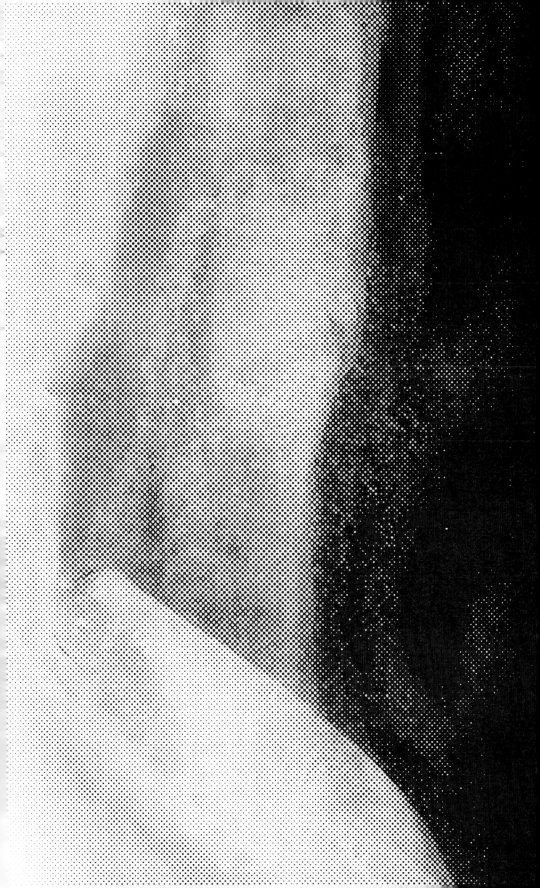

III PRIMARY ABUNDANCE, URBAN PHILOSOPHY—INFORMATION AND THE FORM OF ACTUALITY
VERA BÜHLMANN

I PRELUDE 117 — II ACTUALITY 118 — III CAPACITY 121 — IV ELECTRICITY 126 — V SUN 128 — VI HOUSEHOLDING WITH CULTURE 130 — VII WITHIN THE URBAN 134 — VIII MOTORICS OF SYMBOLS AND ENERGY 136 — IX VALUES 140 — X INVARIANCES 143 — XI MEDIALITY 146 — XII DOUBLE ARTICULATION 148 — XIII CODA 149

VERA BÜHLMANN has studied English language and literature, philosophy and media studies at the University of Zurich, and holds a PhD in media philosophy from the University of Basel. 2003–2007 she held a research position at the Academy of Art and Design in Basel, and since 2008 she has been senior researcher at the Chair for Computer Aided Architectural Design CAAD at the Swiss Federal Institute of Technology (ETH) Zurich, where she has headed the laboratory for applied virtuality since 2010. Her publications include: *inhabiting media, Annäherungen an Herkünfte und Topoi medialer Architektonik*, PhD thesis, University of Basel 2011; *pre-specifics. Some Comparatistic Investigations on Research in Art and Design* (co-editor), jrp Ringier, Zurich 2008; *metaworx. Approaches to Interactivity* (co-editor), Birkhäuser, Basel 2004.

"Either we know what something is, or we do not. If we do, then there is no point in searching for it, if we do not, then we will not know what to search for."[1] [" P.120]

"[...] If knowledge is not to be identified with its object, knowledge is a matter of constructing, using and coordinating symbols."[2] [P.120]

This article argues for a radical perspectivity shift in cogitating the urban, which involves an approach to infrastructures not solely in terms of functionality, but predicated on the pre-modern philosophical terms of capacities and capabilities. Characterizing infrastructures as technological means of maintaining a steady supply of existential basics poorly recognizes the peculiar space of potentiality they maintain and provide along with consolidation of steadiness. The advent of global logistics and media networks not only dramatically enlarged that infrastructurally maintained space of potentiality, but democratized it as well. This space of potentiality is transversal to the nature-culture dichotomy, and can be comprehended as an infrastructural component of urbanity. Thinking in philosophical terms of capacities and capabilities in relation to infrastructures entails the secularization of certain noetic figures related to technics, motion, and power that had a strictly metaphysical connotation while

1 Plato, Meno 80d-2.
2 Ernest Nagel, "Formal Logic and Geometry", in: *Teleology Revisited and Other Essays in the Philosophy and History of Science*. Columbia University Press, New York 1979, p. 255.

they were connected to cosmological or natural frames of reference. This article makes suggestions along the lines of how to conceptualize the triad of information, virtuality, actuality, which today ubiquitously accompanies, and is secularized by, so-called media reality and related technoscience—itself rather urban than natural or cultural—without the triad losing, in the process, the differentiation capacity of its metaphysical past, while being alive to its profanization.

I PRELUDE

"I'm already here," the hedgehogs are calling out to the hare from their distributed places, as he comes tearing down the field like a windstorm. "I'm already here." The hare cannot believe it, 73 times he insists upon making the test, until he falls dead to the ground. This tale of the Brothers Grimm is about a race, provoked by the elegant hare's scornful behaviour towards the hedgehog's clumsiness and his short and crooked legs. As a wager it seems, for one of the parties, to be a foregone conclusion, and nevertheless it cannot play out, because of the other party's cunning contrivance. In calling this wager unfair, one would be jumping hastily to the perspective of moral categories, which to me seems less promising than sticking for the moment to examining the situation "technically". Let us begin by simply looking over the various capabilities that are pitched against one another in this contest. Thus, I should like just to consider the hedgehogs' mental cunning and the motorial nimbleness of the hare, and thus not to go beyond seeing two principles of capabilities in competition that are apparently neither congruent nor comparable against a common metric. Cunningness as a capability grows out of a discontinuous initial situation and construes purposeful continuity through a smart logistic setup, whereas motorial nimbleness as a capability presupposes natural continuity as a steady background. Their apparent irreconcilability opens up a fault-line, which may be seen as separating, as it were, information technology and its operational paradigm in the symbolic, from that older technology that, on the substratum of the physical, represents, processes and transforms continuous synergies and connections.[2] Through the electronically-digitally conveyed nets, general interconnections between information, circulation and organization came to be established in previously unknown fashion, which form the backdrop to the increasingly overpowering logistics in the electronic infrastructures of our everyday life, which were in turn converted into the basis of our urban life over the past 150 years. We experience our urban basis increasingly as relative and heterogeneous, with a multidimensional, net-like structure that above all ought, among other aspects, to be characterized as "social".

II   ACTUALITY

Here, an old question resurfaces in a new context, the one about formal notions regarding possibility and realization. Electricity and IT give rise to the revival of a difficult language game around the linking of form and materiality, through which, since Aristotle and in multiple manners, the notion of "actuality" has been discussed.[3] Aristotle reaches the notion via the assumption of some principle of abstract activity, which he calls ἐνέργεια (enérgeia). It encapsulates the idea of an act that is never-completing and must therefore be thought of as prior to any concreteness in space, time or body. Much of what flows into this figure of thought passes today as metaphysical and unnecessary—even to someone not decidedly thinking of themselves as being positivist. There will be no raising of ghosts here; but we suspect that

---

[2]   For a discussion of this fault-line from a historical perspective cf. Bernhart Siegert, *Die Passage des Digitalen. Zeichenpraktiken der neuzeitlichen Wissenschaften 1500–1900*. Brinkmann & Bose, Berlin 2003.

[3]   The main idea behind any hylomorphism is: "matter provides the potentialities which are actualized by the form", cited in Istvan Bodnar, "Aristotle's Natural Philosophy", in: *The Stanford Encyclopedia of Philosophy* (Spring 2010 Edition), Edward N. Zalta (ed.), http://plato.stanford.edu/archives/spr2010/entries/aristotle-natphil.

it is in this language game, around *enérgeia* and its being-active in an abstract sense, that some index for markings might be found that might allow forward-looking symbolization of the rift that gapes between electronic technics, mechanical technics, and what "technics" itself is taken to mean socially. The central interest of this language game about actuality, from the application-oriented perspective of technics, lies in differentiating between mere ability to do something and being proficient in the ability to do it, in a never-to-be-perfected sense of artifice, which Aristotle, in his natural philosophy, established as a theory of capacities and capabilities.[4] This theory underlies his dynamism of reality, which depends on a relationship between what he called *act* and *potency*, between abstract doing that is never destined for concrete completion, and concrete, corporeal action that accomplishes and completes. In a nutshell, Aristotle cogitated on the character of reality and asked about its beginning, ἀρχή (*arché*). In his explanation of how knowledge about a dynamic notion of reality may be gained, he distinguishes, on the part of potency, the connections of concrete motion, which is liable to being analysed, studied and trained. On the other hand, on the part of the act, he postulates the principle of some *primus movens*, of a mover to whose instigation motion can be traced back in order to be analysable, but about whom—at least within Aristotelian philosophy—nothing further can be said. This moment of an abstract act, from which concrete movement is being triggered, underlies the present re-familiarization of the idea of "actuality", related to what has more recently turned into a somewhat diffusely laden buzzword "virtuality". It would seem sensible to assume modi of actualization liable to be differentiated further than those between possibility and reality; this text would even like to consider the question of whether it might make sense not only to speak of modi of actualization, but indeed to achieve altogether a novel way of accounting for the process of relating form and actuality.

The revival of this difficult language game of actuality relating to the interplay of form and materiality is grounded in the problems that originate from the quest for a categorical concept of information. The mathematician, Norbert Wiener, was probably one of the first to have remarked that information is not adequately accommodated by the two traditional scientific categories of mass and energy; it is reducible to neither category, behaving as it does in a fashion transversal to both. The problems thus arising are so disturbingly unsolved to this day, that it seems meanwhile no longer admissible simply to look at

---

4   Cf. for an introduction to Aristotelian thinking particularly from this perspective: Ludger Jansen, *Tun und Können. Ein systematischer Kommentar zu Aristoteles' Theorie der Vermögen im neunten Buch der "Metaphysik"*. Dr. Hänsel-Hohenhausen, Deutsche Bibliothek der Wissenschaften, Frankfurt a.M. 2002.

information as information, without first specifying whether one's interest be technical, language-philosophical, or indeed one of science history. It is therefore all the more remarkable that just now, in the spring of 2011, a first a-disciplinary and popular-scientific book should show up, under the title *The Information*—note the definite article.[5] The author, James Gleick, said on his blog that when asked by *Wired* magazine for a brief definition of information, he was then short of an answer. On second thought, however, he might submit, "information is how we know".[6]

It is easy to see there a relation, probably more intuitively sensed than dedicatedly stated, between the information question and a philosophy of capacities and capabilities. A relation, however, that appears even clearer in the work of Gregory Bateson who, along with Wiener, is one of the early protagonists to involve themselves in questions about the essence of technically treatable and content-devoid information. He came up with the by now legendary, but also somewhat conjuring, formulation—much more Aristotelian than he might probably have realized himself—that "information is a difference that makes a difference."[7]

Thus Bateson quite openly described information as that red-hot iron which nobody would touch directly, since it re-uncovers the very problems one had thought had gone away thanks to the modern matter-as-mass idea. To define information as something being that is, in its being, pure doing, implies—as does the just discussed model underpinning Wiener's noetic figure—a philosophy oriented towards capabilities and capacities from which modern science believed itself long since emancipated.

What this text proposes, then, turns upon the possibility of a philosophy that is on the line not of a capacity / capability-oriented study of nature, but of information. I should like to proffer such a philosophy as a genuinely urban architectonics. It ought to provide orientation for thought enabling it to deal with the polymorphy of urban actuality. Architectonic considerations are inseparable from categorial determinations that organize the structuring and sorting of what happens. If these categorial determinations are to be capable of dealing with the polymorphy of urban actuality, we need to assume an interplay between them, which we shall call motorics of the urban. This does not imply the city as a constructible machine, nor as an organism with natural aptitudes or capacities, as it were. The point is indeed to find a concept capable of abstracting from both of these ideas. Below, some approaches to this end will be presented in concise form.

5   James Gleick, *The Information: A History, a Theory, a Flood*. Pantheon Books, New York 2011.
6   James Gleick's blog, *Bits in the Ether*, http://around.com.
7   Gregory Bateson, *Ökologie des Geistes. Anthropologische, psychologische, biologische und epistemologische Perspektiven*. Suhrkamp, Frankfurt a.M. 1999 / 1981, p. 582.

III CAPACITY

An urban architectonics presupposes interdependency not only between knowledge and reality, by τέχνη *(techné)* in the sense of artifice, but also between knowledge and technics. Let me begin by taking up a noetic figure of Michel Serres, who sees technics as forever bringing energy stores, and the forces to be culled from them, into constellations. In his *Motors: Preliminary Considerations Regarding a General Theory of Systems*,[8] Serres postulates an indissoluble relation between what people can know at a certain time, and what they are at the same time capable of doing, not only according to an individual's natural capacities but with the population's technical support available at that time. He articulates a generational history of systematic thinking and doing, which he derives from the differing ways in which people have symbolized, at different times, an apparently never entirely positivizable openness of their ability to make the most of their capacities. Serres perceives technics as artifice in the old Aristotelian sense, but also as the realized and concrete architecture of such systemic thinking and doing. Since Lambert and Kant, philosophical architectonics has explicitly denoted the building of systems as an art, within a study of transcendental methods.[9] Demarcated from it, Serres' proposal points in the direction of some meso-architectonics,[10] he proposes including the realized and concrete architecture of systemic thinking and doing into his notion of knowledge, which he understands, with Aristotle, as the characterization of reality. He thereby relates technics directly to something that was divine to Aristotle, and counted as sublime art for Kant: the ability to cogitate systems in general. With this proposal, Serres secularizes that mythical Aristotelian motive of transcendental *enérgeia* to which, in Aristotle, every δύναμις *(dýnamis*, potency) must be traced back. Serres will not consider technics independently of where from and how its actuation is thought to be organized. Thus, to Serres, the difference between the wherefrom-driven and how-functioning is sine-qua-nonical

---

8   Michel Serres, "Motoren. Vorüberlegungen zu einer allgemeinen Theorie der Systeme", in: *Hermes IV. Verteilung*. Merve, Berlin 1992, pp. 43–91.
9   Immanuel Kant, "Die Architektonik der Reinen Vernunft", in: *Werke in zwölf Bänden*, vol. 4, Frankfurt a.M. 1977, pp. 695–709. What is less well-known today is that some years before Kant, Lambert had already published his book on an architectonics of thought. Cf. Johann Heinrich Lambert, *Anlage zur Architektonik oder Theorie des Einfachen und Ersten in der philosophischen und mathematischen Erkenntnis*. 2 vols, 1771.
11 We call it "meso-architectonics" and not "meta-architectonics" because the Lambertian and Kantian notion of architectonics has the transcendental integrated into its methodology. The suggested noetic figure here does not try to further abstract from the assumption of transcendentality, but rather tries to differentiate it. Therefore we call it "meso"—for Greek μέσος *(mésos)*, meaning "middle" or "medial". We cannot elaborate further on this here, but I have developed some approaching thoughts in this direction, in: Vera Bühlmann, *inhabiting media. Annäherungen an Herkünfte und Topoi medialer Architektonik*, 2011, PhD thesis, University of Basel, Faculty of Humanities, available online at: http://edoc.unibas.ch.

to his *Preliminary Considerations*. Therefore, he proposes mutually to interrelate technics and knowledge, without however relating them jointly to a common anchor point, such as nature or culture. I will first follow his proposal of a generational history of systematic thinking and doing, up to a point, so as to carry then the noetic figure somewhat further, in an interpretation of my own.

A first gestalt of such architecture-turned-symbolization of the described difference is found by Serres in the language game of the machine. Machine as a notion harks back to the Greek *makhaná* (means, tool), which in turn traces back to the proto-Indo-European *maghana*, meaning "that which enables". With Serres, we shall associate the machine with technics for which energy stores are latent in nature, e.g. levers, winches, pulleys, mills, turntables. The principle underlying these machines is the linkage of geometrically continuous circular movements, into which ingathered forces are being integrated, and used for driving some further movement. They are all rotative machines. As such they are strictly geometrical, their elements are universal forms, and their enabled motion is mostly used for transport. The technical momentum in these machines is being harnessed through a fixed point, such as the one from which Archimedes famously thought he could move the world. These machines work with motion. In their case, the wherefrom-driven is distributed in natural fashion throughout the world. Machines take energy from nature; they are literally in sync with what happens, to whose "coming-about" they contribute. They convert energy from one form to another, so as to move and transport objects. Michel Serres calls these machines "epistemic machines",[11] because their premise is one point in the without, the standing above happenings, as per the Greek ἐπί *(epí)* for "above, nearby" and ἵστασθαι *(hístasthai)* for "to stand".

A second gestalt of such architecture-turned-symbolization of the difference between a wherefrom-driven and a how-functioning may, with Serres, be identified based upon the treatment of heat. In contrast to machine kinetics, which moves and transports objects, thermodynamics affects matter in its composition. Heat is a uniform principle capable of affecting everything. Heat technology does not transport things, it transforms matter. Following a uniform principle, it effects transformations, it "realizes" states of substances. The epistemic principle of a fixed point in the without dissolves here into an operational difference within the world: heat engines live upon the temperature difference between two sources. Unlike that gestalt of technics that Serres describes as machines, heat technology does not follow any principle of continuous movement deriving from natural distribution, but it does

---

12   Serres, *Vorüberlegungen*, ibid. p. 50ff.

encapsulate such a distributory principle in construed fashion. Heat technology works with the next derivative of continuous movement, in other words temperature, which forms through differing velocity of molecules. We follow Serres' suggestion in calling this technology "apparatus", from the Latin *apparare* containing *ad* for "towards" and *parare* for "to fit out". Apparatus do not give off energetic power directly from gathered energy, but produce an energetic system that gives off power in a steady flow, and, as drive units, work in the running of machines much more efficiently and controllably. For Michel Serres, this interplay between apparatus and machine represents the general characteristics of a motor. A motor differentiates the continuous motion of machines and thus produces not only output, but output capacity. In apparatus, the fixed point in the without of epistemic machines turns into a motor, and Serres terms the knowledge he associates with it "diasteme",[12] from Greek διά *(diá)* for "through, throughout", related to δύο *(dýo)*, δίς *(dís)*, and Latin *duo, bis*, with the root denoting "two".

Let us now carry this noetic figure further, in our own interpretation. The capacity of information technology is fully absorbed neither by its machine nor its apparatus quality. The cybernetic and established language game would suggest that it controls the apparatus that power the machines. But that would ignore the difference between the wherefrom-driven and the how-functioning that we find so interesting in Serres' thinking about general systems. Let us first turn to the occurrence and the architecture in which this surmised new gestalt of technics today manifests itself. At first we shall leave aside "the computer" and for a moment consider it as a conceptual hubris having hypostasized into a universal medium. Then we find information technology in the multitudinous applications as electrical devices, and in the thousands of applications that have invaded electronic networks and propose various services. Apple's App Store opened in 2008, with just over 500 downloadable applications. One year later, 65,000 apps were on offer, and by July 2010, more than 250,000 such apps were available in the App Store alone.

Meanwhile, the infrastructures for supplying motive power too have differentiated in two directions: central and heavy mammoth engines or power plants, and comparably filigreed electronic networks in which motive power is, or at least tends to be, ubiquitous and persistently available.

IT-based applications are powered by steadily available electricity. They are no longer translocation-based as were the first-generation machines, nor based upon orchestrating particle velocity, as were the second-generation apparatus. Whereas the first two generations

---

13   Serres, *Vorüberlegungen*, ibid. p. 50ff.

were both directed at tickling out, in one way or another, the specific capacities and potentials proper to things, it is here much less clear what potentials are to be tickled out, and whether these potentials may be assumed to have pre-existed in latent form previously to being addressed. It feels insufficient for their characterization to say that digital IT apps are powered primarily and directly by available electricity. It might be more accurate to say that they are propelled by the fantasy of their developers, and also that they are aimed at propelling the fantasy of their users. Digital IT apps all offer functionalities whose "purposes" or "usages" are at times apt to hang at stratospheric heights above the manifold facets of our everyday life. The developers of such technology push new modes for everyday dealings and doings, well before any convention for their "assessment" might "situationally" have had a chance to emerge. The driving force behind such apps are developers' phantasmata, who will not and cannot be aware of any actual purpose or usefulness of their brain-children. More specifically, IT developers are not into actualizing things in the sense machine users are, who tickle out the least potency of a thing, nor do they seize upon a thing's natural dispositions for transforming its characteristics, as the apparatus operators do.[13] The IT developers are enablers of genuinely cultural behavioural forms on behalf of urban everyday life, whereby such enabling is inseparable from a process of evaluating the offered apps themselves, an ongoing process all the while that the apps are being accepted, refined, neglected, used, and thus symbolized in their purposefulness. The significance of the things being enabled gets established only through—and dependent upon—the popularity, and actual activity, of using them, and upon the popularity of the modes of applying them.

Against this backdrop, the question about where to look for the specifics of this newer and third gestalt of technics may now be somewhat circumscribed. Serres' noetic figure distinguishes their characteristics according to their wherefrom-driven, or how-functioning. If we are to extend the figure to cover IT as well, account must also be taken of what they are powering. Apparatus in his dramaturgy demarcate themselves from machines by producing not output through

14 Thinking towards "an urbanization of the assumption of transcendentality" that our meso-architectonic theory suggests, relates of course rather closely to many issues debated more recently under the caption of *cognitive capitalism*. My stand here might seem rather uncritical by comparison. Yet the assumption of primary abundance [see pp. 135ff.], and my interest in a philosophy and methodology starting from it, makes the relation to any materialist position complicated and difficult. There cannot be, strictly speaking, something like consumer culture any more—for how can we define waste if resources are not scarce? Instead of elaborating this further here, I would refer to an article the mood of which I share: "Art, Criticism and Laughter: Terry Eagleton on *Æsthetics*", paper delivered at the conference *Æsthetics, Gender, Nation*, organized by the Raymond Williams Trust, Oxford, March 1998, http://www.bbk.ac.uk/english/skc/artlaugh.htm (15 June 2011).

motion directly, but output capacity for motion, through transformation. Let us suggest taking the electric, and digitally contrived, IT applications as drivers for our own phantasmatics, with a view to the development of abilities within apparatus and machine operation. They work in the third derivative, as it were, of motion (of things) and velocity (of molecules). They allow us to put the capacities of a multitude of electrical-device mini-motors into various relations. All of these mini-motors are dissimilarly timed in the heads of the multitudinous urban beings. Putting it a bit differently, they are not facilitating, as the machines are, knowledge in the sense of regular and predictable development along the course of things. Nor do they realize knowledge, as apparatus do, for the optimizing of machine knowledge. IT applications start narrating *with* knowledge, by allowing for constellations of sequences, series, and their integrated embeddings into technically based dispositions, which are therefore in this sense real and not purely fictitious, inasmuch as these narratives are distributed in the populations and infrastructurally supported by electronic applications and devices. IT applications provide integratability of individual phantasmata into socially performed rationality, in the old sense of the Latin word *ratio* for "reckoning assessment", and "establishment of proportionalities".

To address this specific potential, it is not enough to move from the epistemics of machines to the diastemics of apparatus; it takes another step, of cogitating about "knowledge" that were able to meet this new gestalt of technics in IT applications and their electronic devices on an equal footing. Such cogitations about knowledge we will term "choreostemics", from the Greek *choreía* for "dancing, (round) dance", referring to an unfixed point loosely moving within an occurring choreography, but without being orchestrated prior to and independently of such occurrence. The point in the without of first-generation technics was turned into a motor by second-generation technics. This may be associated with the shift from a transcendent notion of knowledge to a transcendental one. Now, we may take this evolution one step further. Along the way, philosophical thought has freed itself from the primary assumption of to-be-referred-to identities, having started to view these identities as differences. Choreostemics lets these differences be treated as operative differentials. It stands for knowledge about the motive dynamics within thinking. Its point is no longer knowledge accumulation or production as is that of diastemics or epistemics. Choreostemics is about training a capability of differentiated behaviour within produced and accumulated knowledge.

To carry Serres' generational model further, the two gestalts of machines and apparatus are to be joined by a third, that of applications. These generate and maintain, on the basis of epistemic knowledge, a diastemic space for dealing with such knowledge. However, they shift orientation of such dealing with knowledge to the social domain, depriving it of its natural anchor point. Said more graphically, everything we have so far sensibly or non-sensibly stated about, or ascribed to, the world, is apt to be symbolically coded and processed as information. This processing sphere also contains the very properties of the physis that is describable through the scientific categories of mass and energy. These properties may themselves be symbolically coded on an atomic level. Furthermore, on the same level and in the same format as the respective coded knowledge about these properties, the properties themselves may be "talked" about as information. Thus, we may print out, in almost unlimited quantity, instances of a combinatorially arranged physis, and compose and configure their properties; when networked, they may form entire landscapes. These landscapes include public utilities, waste disposal, transit systems, goods logistics, telephony, GPS sets, and social media such as Facebook. They develop from electrically interlinked logistics turned infrastructure, and thus are, in a traditional sense, neither ideal nor concrete, neither neutral nor territorial, neither artificial nor cultural.

We shall call them "culturly" ("kultürlich" in German, a combination of "*kult*urell" and "nat*ürlich*") or "urban". They are landscapes of possibility, attainability, accessibility, applicability, tradability, in short, of actualizability. This third generation of technics evolves from electricity as its substrate. Turning now to this electricity-as-substrate, we propound the distribution of technical-energetic instances being brought into wider relations than just that of the topology of installed power grids. The networked distribution of such technical-energetic instances form out into culturly urban landscapes, charged with potential, and culturly urban loci formed from the manifold ways of how this potential actualizes and even materializes. We are now out to comprehend these strangely distributed loci in their dynamics, which are, for a substantial part, driven by the immanent modes of dealing with them and their accommodative culturly urban landscapes. From these dynamics results the heterogeneous structure of our present, increasingly urban, life.

IV   ELECTRICITY

Under the aspect of their formality, these culturly urban loci may well be perceived as abstract in a sense that is comparable to how electricity, seen under the aspect of its formality, may be perceived as abstract. Any concrete loci may be instantiated from this culturliness, as may any form of energy from electricity. Since the closing years of the 19th century, we have experienced no less than an "information-technical" development, to which there is perhaps only one parallel. While Socrates may be credited with initiating that "speech-technical" development that "brought philosophy from heaven to earth",[14] what we have lived since the end of the 19th century may well some day prove similarly momentous.

In Greek antiquity, the combination of enlightenment and city-state, of free speech and its cultivation through phonetic writing and rhetoric, helped to create the foundations for our ongoing understanding of science and philosophy. As demonstrated by the extraordinary importance taken by earth mensuration as well as the *mos geometricus* within every concept of knowledge—both epistemic and diastemic, in our terminology of generational history of knowledge and technics—territorial ordering and structuring, in terms of quantity, scope, and the systematic proportions between them, are basic to the "earthed" referential relations that spring from them. In the same vein, Plato accepted only those into his Academy that were geometry-literate, and Aristotle oriented his invention of systematic treatment of statements in syllogistics based on the way of treating geometrical elements systematically.[15] IT, on the other hand, accelerates a process today that might, somewhat pictorially, be described as vaporizing our semiotic grounds, and our knowledge derived from them, into some purely formally treatable symbolic.[16] Digital codability causes volatilizing of what had been earthed and solid, into a formal "symbolicness" that is not directly lodged in earth, or things, but primarily in the abstract element of the electrical. This has impacts on how we conceive of the composition of reality. Within the electrical, symbols may be treated purely formally, following concept or experiment, within the already comprehended and without, fantastically, and may be printed out and reproduced into any composition of mass, energy, or signs.

This abstractness of the electrical found early notice in scientific discourse,[17] and somewhat later in more popular culture as well—Dick Raaymakers, e.g. a pioneer of electro-acoustic music, published in 1979 a manifesto under the title *The Art of Reading Machines*,

---

14   Cicero, *Gespräche in Tusculum*. Translated into German by Ernst A. Kirfel, Reclam Verlag, Ditzingen 1997, 5.4.10.
15   Already in antiquity was there a vast dispute around the relation between geometry and arithmetics, relating to the notion of finitude and delineation. A science of physics needs to defend itself against the assumption of infinity in order to be systematic. In order to get a general idea of how in antiquity, the comprehension of a finite cosmos was sought after, cf. the short text by Archimedes: "Über schwimmende Körper und die Sandzahl", in: *Ostwalds Klassiker der exakten Wissenschaften*, 213, Leipzig 1925.
16   For an overview of the problematics of an algebraic (symbolic) treatment of quantities, cf.: Augustus de Morgan, *The Connexion of Number and Magnitude: An Attempt to Explain the Fifth Book of Euclid*. Kessinger Publishing, Whitefish MT 2009.
17   Especially for the discourse around the theory of relativity in the first half of the 20th century, cf.: Albert Einstein, "On the Electrodynamics of Moving Bodies". In: *Annalen der Physik und Chemie*, 17, 1905, pp. 891–921; Herrmann Minkowski, "Raum und Zeit". *Jahresberichte der Deutschen Mathematiker-Vereinigung*, Leipzig 1909. The postulates of special relativity are undisputed today and have, e.g. via quantum electrodynamics, become the ordinary fundamentals in the design of computer chips.

where he mentions a crucial observation he made during his work for Royal Philips Electronics Ltd: "The electrical device essentially differs from the mechanical in that its components do not move."[18] The power-productive principle of early heat machines, as well as of any mechanical implement in general, lies in the movement of its components. "Movement implies freeing oneself from the ground in some way or other"; in mechanical machines, their functioning can be traced to their positioning and the circumstances resulting from it. The synergy of the positioned components is orchestrated via material contact, or, putting it differently, through transfer of energy. Very differently in electrical machines: "When an electrical device is functioning, its insides are charged." Energy transfer is being replaced by energizing or charging with energy. The electrical implement does not function through some link to the ground, but through being disengaged from it. Unlike mechanical machines—the dynamics of which are also due to being "freed" from embedment in the ground, yet constituted just through this reference to the ground—electrical machines pull back from the ground into a shell, and form their own energetic compartment, as Raaymakers goes on to say: "This means the inside, through its electrical charge, disengages itself from the earth and out of its housing [...]."[19] The principle of power production from electricity is still none other than that of moving components—electro-chemically, electrostatically, or electro-dynamically; but the principle of encapsulation with regard to the electro-technical does not follow from this role held by movement in classical mechanics.

Even if nothing moves, the electrical current is still present as a driving potential. It moves at 200,000 km/s through grid topology, feeds by now a worldwide technical population of 500 billion electronic implements, and has no doubt become foundational to 21st-century urbanity.

V   SUN

Yet one may possibly be left, as Raaymakers also was, with some unease about one's own fascination with electricity, in case one perceives it as "material", or as "form" of energy. From such a perspective, he noted, "Electricity is the poorest and most exploited of all matter." There is a concrete reason for our insistence upon divorcing electricity from the perspective of materialistic thinking, and perceiving the potentiality of energy in its abstract formality in it. When pursuing this train of thought—admittedly a complicated one—with a mix of openness, scepticism, and pragmatic expertise similar to Raaymakers' in his

18   Arien Mulder, Joke Brouwer (eds), *Dick Raaymakers, A Monograph*. V2_ Publishing, Rotterdam 2008 p. 10 [134].
19   *Raaymakers*, ibid., p. 10 [134].

manifesto, it will lead us in a twofold way towards a modified relation between symbols and energy, or, more familiarly, between what we commonly distinguish today as the key elements of culture and nature. With respect to this distinction, the sun plays, in ever-changing fashion—but at least since Plato and throughout history—a central role in philosophy. Domestication of electricity, as an abstract, formal "recipient" for various forms of energy, also affects the meaning of the sun for our concepts of nature and culture. This has long since become apparent in our talk about artefacts and a kind of artificiality for which we lack philosophical categories that would let us put them in relation with things natural. Perhaps the most crucial thing about artefacts is that the method by which they are being produced has ceased to be a reference for what they mean to us. This referential relation was still extant and relatively unequivocally constituent for the crafts, and for manufacturing, such as in the way smithcraft clearly calls metal to mind. Whereas this unambiguity has long since been cancelled out by technical production and duplicating processes, such decoupling between product material, structure, and capacity, is being taken a radical step further by the electronic printing methods—in this respect, digital printing processes are just indices pointing to a mere formality manifesting itself in a diversity of materials and structures. The products may thus, on an energetic-atomic level, be equipped with capacities or capabilities that not only do not refer to any specific materiality any more, but that are no longer direct references to any materiality whatsoever. The formality from which printing happens consists of digital code, and must first be evaluated through interpretation and then brought into continuous form.

All this quite touches upon the sun's central meaning for all the language games assuming that in the sun's light and its clarity our concepts of knowledge and insight may be organized. However, there is another—and so far much less pondered—aspect that more importantly wants looking into. For the first time in history, thanks to photovoltaics, we are given the possibility of collecting and storing energy from sunlight directly.[20]

---

20 It has of course been possible to use glass for concentrating the sunlight enough to spark a fire, for example, as Archimedes is famously said to have set Roman ships on fire with the help of parabolic mirrors. But thus only a direct effect may be achieved; storing energy from sunlight became possible only with the latter's direct conversion into electricity. And still today, photovoltaics as energy technology is said to be hampered by the lack of adequate storage devices in the form of batteries; however, such difficulties seem not to be of a principle order, but related to the assumption that we depend upon exploitation of scarce energy resources. That assumption discredits any procedure whose efficiency rate is not on the high side. It has long been known, for example, how gasoline may be produced through artificial photosynthesis, literally from electricity, water and air. Audi, the German carmaker, has recently hit upon this as an opportunity for using excess energy produced by wind power—which troubles electricity grids—for producing CO_2-neutral gas fuel: cf. http://www.solar-fuel.net (16 June 2011).

The sun permeates our living environment, which we have come to appreciate as nature. This permeation exceeds, not just metaphorically but quite measurably, the quantity of solar energy that is encapsulated within nature itself, and stored in nature's fossil, biophysical and biochemical compartments. Every day, solar radiation transfers to earth 10,000 times humankind's daily energy consumption. By taming electricity, we are for the first time able technically to harvest energy from solar power directly, and not indirectly by resorting to natural resources. Thus, we become capable of encapsulating energy in a genuinely culturly urban sense, and of integrating it into our environments—quite in addition to the solar energy, as whose storage and transformation system we have through science learned to see nature.

Now, this technology possibility has not been with us since just yesterday, but for about the last hundred years. What changed with the more recent information technology, however, significantly affects the production economics of this kind of technics. Photovoltaics can be produced by printing processes and thus is the sole energy technology whose development follows economically the lines of IT: symbolic imprinting at the atomic level, for producing printable solar foils, is likely to be the only technology that will become not dearer over time, but cheaper. Massively cheaper, such as the computer chips over the past decades. In the photovoltaics area, the products of literally symbolic imprinting at the atomic level suggest the evolvement of a genuinely culturly urban energy household.[21] Thanks to solar foils, energy turns into a consumer product. Not in spite but because of this, photovoltaics today represents a very realistic infrastructural option for energy production,[22] however challenging, in terms of sustainability within an overall energy balance, this may be for materialist ethics, which is predicated upon the distinction of *consumption* from *work*.

VI   HOUSEHOLDING WITH CULTURE

Photovoltaics is not only a realistic infrastructural option; in its wake new philosophical questions crop up, which revolve around ideas of abundance. Whereas the idea of abundance has never played a notable role in natural philosophies, it has appeared in economic theory at least since Marx. The acceptance of abundance, however, with which

---

21   Cf. Ludger Hovestadt, Vera Bühlmann, *The Power Path. A radical pathway from energy crisis to energy culture.* Forthcoming 2011/12.

22   "Solar panels are coming down dramatically in cost per watt. And as a result of that, the total amount of solar energy is growing, not linearly, but exponentially. It's doubling every 2 years and has been for 20 years. And again, it's a very smooth curve. There's all these arguments, subsidies and political battles and companies going bankrupt, they're raising billions of dollars, but behind all that chaos is this very smooth progression," says the futurist Ray Kurzweil in a recent interview entitled "Solar will power the world in 16 years". http://bigthink.com/ideas/31635 (16 June 2011).

we are dealing here, refers to it as something primary, and ought to be distinguished from any notion of affluence, excess or surplus. Unlike these, abundance is not the outgrowth of a temporary feature of some system benefit—which in a finite system must occur at the expense of another sector. The notion of primary abundance, however, considers abundance as fundamentally indefinite.

Primary abundance is indefinite in the sense that it is a potential not yet rated, which is not the same as being without value or worthless. In this, we see the reference plane for a choreostemic adaptation of the philosophical notion of virtuality. Perhaps this indefinite potential may be described as a potential not endowed with one specific form but arising in the appearance of literally any gestalt. In this very quality, in the actualization of the formality in which it will make its appearance, it seems as yet to be unspecific. Physically, electricity itself may be considered as a formal potential that in random forms can become energy, and, via energy, power.[23] Thus, we are beginning to trace the connection between value and form. "Indefinite" literally means without bound, "undelimited". Undelimited means that the potential is not yet to be brought into continuity. Normally, this is just what forms and values do. As intermediary principles, they move formless potentials into continuity; but they do it traditionally in multiple spheres: values are for calculating and householding, with respect to the stock that may, in "earthed" fashion, be demonstrated, described, represented, secured, and therefore also construed by means of the forms. Forms are, already in other sign practices than that of the digital code, symbolically put up, and detached from their territoriality or materiality.[24] Within the digital code and electricity combination, however, they may moreover be

---

23 In this context, it is important to define "formal potential" explicitly as it differs from some other concepts that might easily be confused with it. First, we do not mean that which is commonly referred to as "potential energy". In physics, potential energy is a complement to kinetic energy, and in that combination led to the physical measure of *work*, related to the conservation of a momentum or force, which as a measure proved to be much more useful than Leibniz' initial assumption to the same end, that of a *vis viva*. By mentioning here electricity's potential of being converted into any form of energy, we are not calling either for a revival of the concept of *vis viva*, such as Helmholtz was tempted to do when wondering about the electromotive force in the 19th century. Our interest here is not primarily the physical *conservation* of specific potential, but formal *conversion* of the pre-specific potential. The assumption of primary abundance allows us to make this distinction. The indefinite potential we have in mind for electricity, is a purely *formal* potential and not in need of a physical constant behind it; this is also where it differs from Daniel Bernoulli's principle of *Virtual Work* and from the *d'Alembert Principle*, which together form the core of analytical mechanics worked out by Lagrange—to which we will turn in a moment. For a historical overview of the evolution and interconnections of these concepts, cf.: Jennifer Coopersmith, *Energy, the Subtle Concept. The Discovery of Feynman's blocks from Leibniz to Einstein*. Oxford University Press, Oxford 2011, especially the chapter "A Hundred and One Years of Mechanics: Newton to Lagrange", pp. 91–147.

24 We are thinking here of sign practices in general that raise the claim of systematic formal coherence, from a mutually stabilizing interrelation between arithmetic, geometry, and language, as opposed to dealing with signs in terms of open lists, for example, or inconsistent or incompletable tables.

energetically charged and thereby "bodily" articulated, as it were. This takes place without possibility of securing, in unambiguous, i.e. binding fashion, the anchoring of forms in some reality-related physical. The forms with which we computationally operate can therefore no longer be regarded as naturally "motivated" or "conditioned". Consequently, the bodies created from these forms tumble, as it were, out of the orders perceived as being natural. Putting it differently, their state becomes problematic—today we speak of artefacts in this respect.

This arbitrariness regarding motivation and conditionality of forms is a subject that in 20th-century arts has long since begun to be differentiated, in numerous discussions about abstract forms in painting, open forms in music, in cinematic art about Jean-Luc Godard's thinking forms[25] and Peter Greenaway's distributed forms,[26] as well as in the predominant theme of form-finding in architecture, or in design more generally. In computer science, the unconditionality of forms has indeed become quasi-natural, without, however, being adequately comprehended and discussed in the implications that touch upon the comprehensive topos of "calculability". This arbitrary form-related motivation has by now seeped into everyday life as a matter of course. Today, everybody can do many things. And not only that: everybody can move and initiate a lot. Or, with a pinch of drama, it almost appears as if, against the backdrop of these developments, Aristotle's first cause, that "instance" of *enérgeia*, had, through the millennia but since the taming of electricity at tremendous speed, exteriorized into some kind of "instantaneous logistics". At present, everything believed and known tends to be digitally reconditioned and made accessible as information, for general application and playing-out. Given access to such culturly urban infrastructure, anybody, according to circumstances, may be author, mover, initiator, transformer, educator etc. Regarding such applicability itself, it is unproblematic that, in the information format, knowledge (e.g. with regard to nature) can no longer be easily sorted out from the flickering distortions of the unreliable and its fallacious appearances. In the urban household, values and products are being created from either of them. Today, causes themselves, as intermediary "instances" between form and matter, cannot any longer be adequately separated into being either "metaphysical" or "natural". Those two language games clearly seem insufficiently differentiated for addressing the culturly urban consistency of things, whereby culturly urban consistency means "phenomena caused out of this primary abundance of potential that is as yet without concrete form". Such

---

25 Jean-Luc Godard, *Histoire(s) du Cinema.* 1998.
26 Peter Greenaway, especially: *The Tulse Luper Suitcases,* 2004, consisting of different films, exhibitions, books, websites, among other formats. In this project, Greenaway presents a history of the 20th century as a biography of Uranium, which of course can only appear in a multitude of forms and formats (this, at least, is one of Greenaway's narratives around this project).

phenomena "are" in genuinely pre-specific fashion, pre-specific to mean neither discontinuous nor continuous. Rather, for such culturly urban consistencies formed out of primary abundance, the only characteristic may be seen in the fact that they are prompted and caused from some exuberance; as phenomena they always forerun the process of their becoming assessable and significant, and are in this sense pre-specific. Above, we referred to this formless, not-yet-assessable potential as virtuality. In parallel to this runs what we have described a choreostemic turning-about from technics as a swing from apparatus and their physical motricity towards applications in and from electronic-logistics networks, the "motricity" of which we described as culturly urban.

It is probably part of the timeless invariance of technics, to push human ability to do into some not entirely positivizable openness or artifice. The attempts at a new symbolization of this incompletability are visible throughout the 20th century.[27] In the modern legacy of "earthed" thinking, the distinction between that which was "realistically" conceivable and that which was to be seen as pure fantasizing, was a central concern of symbolizing this incompletability. The orientation of values took place around the topos of feasibility. Today, a new symbolization of incompletability is about to appear of a kind that relates to the conceivable itself, and the limitedness of our cognitive capacities. Of this, 20th-century philosophical milestones are harbingers, such as Husserl's *"Crisis"*, Wittgenstein's *"Tractatus"*, or Gödel's incompleteness theorems, as well as more recent approaches and their techno-scientific paradigms in brain research on the one hand and fundamental physics on the other, with its attempts at "positive" proof of the Grand Unified Theory of Everything. But as this very example nicely demonstrates, even if stabilizing the feasibility topos as an anchor point is at stake, success is not achievable otherwise than by being geared towards the conceivable for orientation.

The real challenge for householding with culturly urban consistencies in an urban philosophy lies in the fact that "budgeting" them "can" no longer originate in circumstances of natural shortage, but "must" be organized from as-yet-unspecific potential-related abundance. For quite some time now, economic and legal developments, such as licensing, franchising, Open Source, or Creative Commons, as well as the technical vocabulary for their organization, have pointed to the inadequacy of our culture-philosophical notions concerning labour, property and wealth. To a large extent, these go on being locked into

---

27 The turn to such incompletability shows up in Cassirer's *Symbolic Forms* as well as in the turn to phenomenology and existential philosophy since Husserl and Heidegger, and also within analytical philosophy and its search of delineating realms for legitimate statements about the world. As a topos of its own, this incompletability has been popularized by post-structural and post-modern theories, and more specifically in the schools of cybernetic and system theory and its many currents, all inspired by Cantor's noetic figure of ordinal nesting, which organizes cardinal distributions.

traditional differentiation, between authenticity, in the sense of "Eigentlichkeit" as relating to forms deemed naturally motivated, and inauthenticity or "Uneigentlichkeit" as relating to forms deemed unconditional or arbitrary. This is inadequate for representing and reflecting the problems relative to valuation and investment in the context of culturly urban consistencies. The central role of electronic, IT-related infrastructures of the urban leads to the appeal for an urban philosophy that would be called upon to rethink that relation between "eigentlich" and "uneigentlich", oriented towards a pre-specific, or towards a primary abundance of potential that is not yet encapsulated, or integrated into any form.

We suspect the dynamics between actuality, form, expression, and their bodily manifestations to be at the core of such a philosophy. The capacity for a more differentiating view of such dynamics seems to us of central interest for culture-philosophically embedded, but genuinely urban, "householding"; we suspect it to coincide with the above-mentioned paradigm of choreostemics that is being enabled by today's application techniques and in which the fixed anchor point, or, conversely, the point of initial drive, spreads out as "centrality" across the populations. The experience of such distributed centrality is already commonplace nowadays, and is typically called "mediality". We detect, in the medialization of what to Wittgenstein were still Lebensformen—meaning the "patterns or schemata according to which the members of a community live their lives 'in deed' ["in der Tat"], which give them orientation for leading their lives"[28]—a vaporizing of these Lebensformen into a formality of living conditions. In this medialization dynamics, we witness not only the distribution of the schemata of applicatory performance across the population of the communities, but furthermore also a distribution of that fixed drive and anchor point. As a restless "point", as it were, it spreads across the societal communities and perpetrates its moment, from distribution, again and again in different fashion and different places. In these circumstances, it does not seem exaggerated to say that actuality, or what we are used to describe as such, may happen in different forms. Information, particularly as a technically treatable abstract, plays a constitutive role in these dynamics.

VII WITHIN THE URBAN

Phenomena presenting this status of "Uneigentlichkeit" were of course always part of the supportive structures of common social life. But with electricity and information technology gaining in importance in

---

28 Michael Kober, "Die Funktion des Begriffs der Lebensform bei Wittgenstein", lecture held at the DFG-Rundgespräch "Lebenswelt in Wissenschaft, Ethik und Politik", organized by Felix Mühlhölzer and Julian Nida-Rümelin 11–13 October 2006 at the Carl Friedrich von Siemens Foundation in Munich. The manuscript is available online at: http://www.nida-ruemelin.de/docs/vortr_kober.pdf (16 June 2011).

our everyday life, these phenomena are taking a supportive part on an infrastructural level too. Since the domestication of electricity, technology has been losing its stabilizing self-evidence of functioning-thus-and-not-otherwise, and becomes variably composable out of mediality. This mediality unfolds within scientific, social, political, economic—or initially "simply" within "logistic"—networks.

In urban research, this tendency has long since been recognized and studied. In his *Theory of Good City Form*, for example, Kevin Lynch introduced a differentiated terminology for analysing, describing and modelling urban dynamics across varying scales.[29]

Working with such a terminology however must needs be guided by an idea of "city" and thereby falls—or rather remains—dependent on philosophy. Today's urbanism distinguishes widely between three big normative city models: the city of faith, city as a machine, and the ecological city,[30] in which varying paradigms may easily be recognized that all refer to relations between nature and rationality, or between philosophy and technology, and the resulting different conceptions of science. Regarding their—perhaps factual rather than explicitly claimed—normativity, these city models follow the notion of a determinant, final or initial, state, whether it be the idea of a holy city, a functioning machine, or ecological equilibrium. Short of achieving its own comparatistics regarding these paradigms, urbanism remains, even while including geographic, economic, political and social characterizations, tributary to some unconsidered "Weltanschauungness"—which explicitly shows in the term "urbanism" and its -ism suffix.

Approaches to an urban theory that are less directly application-aimed than urbanism or urban development, but take social sciences as their starting point, often refer to Henri Lefebvre. His assumption of an urban revolution tries indeed,[31] by way of the noetic figure of urban totalization and particularly with his idea of producibility of space, to break free of the idea of an anticipatable final or initial state.[32] But he merely shifts the problem away from model-relatedness into an historical process; his urbanization process develops along a time-line, from agrarian society towards city, via the Greek *polis*, the Roman *urbs*, the mediæval town, to 20th-century total urbanization. Yet, what makes his approach

---

29 Among his concepts for speaking about these topics, one finds for example notions like "galaxies", "polycentric nets", "lacework nets", "alternating nets". Cf. Kevin Lynch, *[A Theory of] Good City Form*. MIT Press, Cambridge, Massachusetts 1981.
30 David Graham Shane, *Recombinant Urbanism. Conceptual Modeling in Architecture, Urbanism and City Theory*. Wiley, Chichester, West Sussex 2007, p. 7.
31 Henri Lefebvre, *La révolution urbaine*. Gallimard, Paris 1970.
32 Henri Lefebvre, *La production de l'espace*. 4th edition. Anthropos, Paris 2000. For an interesting discussion of his theory of the production of space cf. Fernand Matthias Guelf, *"La révolution urbaine", Henri Lefebvres Philosophie der globalen Verstädterung*. PHD thesis, Berlin University of Technology 2010, available online: http://opus.kobv.de/tuberlin/volltexte/2010/2537/ (16 June 2011).

interesting is without doubt his assumption of some spontaneity inherent in the arts, a creative act that is part of what he treats under the term of urban poiesis. Thereby, Lefebvre opens the way for an æsthetic-political practice of action, with Aristotelian poiesis still passing in his theory as appropriation of nature *(physis)*. Nature however—according to his strong thesis of urban totalization—to him is not in the first place related to a natural outside, but to the interior nature of the senses, sensibility, sensuality, needs and wishes. Hence, in his meta-philosophy of the urban,[33] to him all poiesis is creation. He assumes a practically unlimited creative capacity that, in his meta-philosophy, turns into the motor of societal development. His thinking thus develops along the lines of reformulating the Aristotelian relation between act *(enérgeia)* and potency *(dýnamis)*. But he pulls it back into the language game of revolutionary development, and his theory therefore fails to integrate productively the challenges of primary abundance; the motorics of the revolutionary is perforce committed to a logic of scarcity and re-evaluations of shortages. The idea of primary abundance, however, as considered here, pre-exists any "valuableness" and is impervious to re-evaluation processes.

Nevertheless, we are going to take up Lefebvre's move towards cogitating the urban, and its enquiring about a dynamic genetic principle of urbanity. But we will not do this within a model of politico-revolutionary thinking, but by looking more fundamentally into the essence of the relationship between philosophy and city. Our proposition is directed at a philosophical architectonics that eschews attempts at defining *arché* as an historical or quantitative pinpoint, but as a mobile point instead, at once moving and being moved as it were, within architectural motorics. Such motorics will then be the point of reference for the choreostemics that, as we suggested, was to accompany the capacities and the potential of today's electronic application technology. We shall submit a concept that might open a way towards acquiring a differentiating manner of dealing with the situation of primary symbolic-energetic abundance. To that end, our architectonic motorics would have to facilitate a comparatistics of Weltanschauungen that would be of great importance for culturly, urban economy-related assessment processes.

VIII   MOTORICS OF SYMBOLS AND ENERGY

Ever since Aristotle, it has been philosophical architectonics' job to assist in the sorting of ideas, with a view to facilitating reliably conceptions of support, stability and order. Herein lies also the fundamental meaning of geometry and its reliability in applicative constructive

---

33   Henri Lefebvre, *Métaphilosophie*. Editions de Minuit, Paris 1965, here cited from the German translation: *Metaphilosophie*. Suhrkamp, Frankfurt a.M. 1975, p. 43 [orig. p. 50].

thinking. Regarding urban philosophy, it cannot suffice any more to rely upon constructive thinking; indeed, it ought to aim (again) for a more comprehensive concept of cultivation *("Bildung")*. Such philosophy cannot any longer be about house-building, complex-building, or system-building; central place must indeed be taken by ways of structuring and organizing that may be regarded disconnectedly from what is concretely being structured and organized. Figuratively speaking, in an urban environment today it seems somewhat inadequate to say one inhabits a house that may be owned, in the full sense of ownership. Instead, it would be more appropriate to say that, while inhabiting "one's own house", one lives *into* something happening, which was there before one's arrival, and is going to remain after one moves. In this happening of urbanity, philosophy must facilitate notional coordination, and help with learning to differentiate urbanity as such. Its architectonic needs to be able to provide integrity while inhabiting the culturly occurrent.

Agreeing with Lefebvre, we also believe urban thinking must be unshackled from the traditional town / country dichotomy, in which the city had the role of centre of power, order, and society. Architectonic motorics of the urban that, as architectonics, does not aim for construction but more comprehensively at cultivation *("Bildung")*, must cultivate the subtle difference between centralization and mediation, in the sense of tending. Staying for another moment with this technical language game of motorics, let us examine the relationship between a wherefrom-driven and a how-functioning, which had already guided our earlier discourse about technics and knowledge. With urban motorics, this relationship complicates one step further, because now the question about the where is added. A technical engine, we might say colloquially, lives in the world; Lefebvre's revolutionary motorics lives in history; but where are urban motorics to live, such as we attempt cogitating them here?

For our proposition, we shall differentiate the language game of localizability. We phrase the question about the *where*, in accordance with thinking in capacities and capabilities, as a question about the *where-on* and *where-in*. Our proposition of an architectonic motorics of the urban comprehends the electrical as the urban's substrate. Aristotle differentiates his notion of substrate, through a combination of forms, and their dynamics of a fourfold causal theory,[34] into ontologically structurable substances. We would now be foreshortening the issue, were we simply to substitute the notion of information in place of the Aristotelian

---

34 Aristotle, in the second book, third chapter of his *Physics*, distinguishes four causes for a causality theory (material, efficient, formal, and final). These four causes are not to be taken dissociatedly, they "effect together", inseparably. For an introductory overview, cf. the article in the Stanford Encyclopedia of Philosophy: Andrea Falcon, "Aristotle on Causality", in: *The Stanford Encyclopedia of Philosophy (Spring 2011 Edition)*, Edward N. Zalta (ed.), http://plato.stanford.edu/archives/spr2011/entries/aristotle-causality/ (16 June 2011).

forms, and a modern notion of causality in place of his causal theory. This is the very nub around which all problems of a scientific-categorical definition of information turn in circles. Information clashes with both the assumption of universal forms and that of uniform materiality as implied by a modern notion of causality. Let us therefore try to reflect more abstractly about a notional combination that might differentiate the electrical as substrate of an architectonic motorics.

Two main players present themselves for notional-motoric teamwork: symbols and energy. These two pass in different ways—but always in interplay—as being constitutive for the language game about the electrical, as well as those about substrate, and also those about the urban. Let us first look out for similarities. Seen from sufficient distance, both language games, that of symbols and that of energy, seem to be about explicating the possibility of "liability" in dealing with "effectiveness", and admit a shared reference to it. Both deal with apprehending effectiveness, without however sharing similar conditions. In the one case, the liability explicated via the symbol notion is geared around the gravitational centres of "meaning" and "sense", in the energy case around "power" and "drive". The notion of energy traces back to the Aristotelian neologism of *enérgeia*, without however entirely coinciding with it. Aristotle probably derived his notion of *enérgeia* from the Greek ἄγειν *(ágein)*, for "to guide, to lead, to command, to carry, to bring". The Scottish physician, William Rankine, who in the 19th century played an important role in the reintroduction of the energy notion into physics, derives the term from the Greek ἐν *(en)* for "in, inside, within" and ἔργον *(érgon)* for "work, act, action". The introduction of the notion of energy aims, both in Aristotle and in more recent times, at the opening-up of transformative activity, or of the changeability of the transient world, in a systematic-organizing manner. For all the shifts that intervened between that early metaphysical acceptance of the term and, for example, its more recent, thermodynamic one, energy for us today is still a principle of continuity that must underlie the externality of objective things—energy is neither expendable nor producible, and for this reason its notion allows of physically describing, and empirically testing, change.[35]

Relating to our outlook upon architectonic urban motorics, we shall now look for a similar setup with respect to the symbol notion. Here again, our first question is about the purpose of the notion, and not about a specific meaning. Hence, we turn to a pragmatic variant of dealing with symbols. *Symbola* used to be small clay tablets, or wooden sticks that

---

35 For an elaborate overview of the reintroduction of the energy notion in modern science, cf.—although no reference is made to the Aristotelian legacy of the concept—Coopersmith, *Energy*, ibid., especially also chapter 14 entitled "The Mechanical Equivalent of Heat", pp. 246–263.

were broken into pieces for evidencing a legitimate claim for fulfilment of a contract or promise. The term traces back to the Greek σύμβολον (sýmbolon), a nominalized neuter of σύμβολος (sýmbolos) for "coincident, something that meets", related to συμβάλλειν (symbállein) for "to put together, to compare". Putting it abstractly, such *symbola* managed the distribution of effectualities in the social sphere. Over time, there resulted a rich language game in which the current meanings of "sign", "mark", "proof", "contract", "passport", "password" and "code" also play a part. The controllable formal fit among the elements of any symbolics seems to be an invariant of the symbolics notion across time. Individual symbols, by themselves, mean nothing to us—or may, on the other hand, mean quite anything, as any conspiracy plot will illustrate.[36] Symbols must stand in relation to a referential framework, because they gain their significance from a context that must be assumed in order to ascertain the way of dealing with them, of how to "apply" them, as it were. Symbols represent bindingness, needed for orienting the expectation of future developments and for having reasons for counting on anticipating them. Yet, they are not doing this through forms directly; they are achieving a formality by rendering values, formally packaged and "encased", procedural through organizing their formal fit on an arbitrary yet formally coherent basis. Through a distribution principle, they establish a fabric of differential bindingness supported not by firm grounds but by distribution. Within such fabric, proceduralness may emancipate from time-spatial actuality, and be symbolically encapsulated. Of course, this relationship between *symbola* and their referential frameworks has been subject to varied and highly intricate shifts and changes. But even today symbols are being called in—e.g. in mathematical equations—to legitimize or discredit expectations of future developments.

While symbols and energy were facing the separate spheres of nature and culture without being motorically related to each other, they could be treated without their respective symbol-ness and energy-ness ever being thematized. This being so, discontinuities were strictly a matter for one's own brains; in nature, all processes are unquestionably regarded as continuous. If we now try to disengage the thinking about city from the traditional urban-rural dichotomy, it becomes clear that in the urban we are not just dealing with symbols within natural energy-ness, but with symbols within a sphere where symbol-ness and energy-ness intermingle, which we are calling the *symbolic*. Such symbolic, however, cannot be taken as a kind of "nature of the urban", unless remaining entangled in ideological premises. The language games around the notion of nature imply continuity of processes that would be incompatible with the urban values of diversity, freedom of expression, negotiability etc.

---

36 A further aspect related to meanings of symbols becomes apparent here, that of "creed", but without direct relevance here for our subject, as we are interested in the formal aspects.

Let us therefore assume electricity to be a substrate of the urban, and see how far that can get us. To us, however, electricity is not, to be as we have already explained, a form of energy, but a potentiality of energy in its abstract "formalness", which makes electricity into something like an abstract container for *any* possible forms of energy that can be actualized from it.

So let us sum up, and revert to our triple question called upon to give orientation to an architectonic motorics of the urban: what in, whence driven, and how does differentiation of its materialization work? We stick to Serres' idea of relating what can be known at a certain time to what can be done at a certain time, with technics support intrinsic to such doing. Thus, the architectonic motorics does, as we assume, work within and upon the substrate of the electric. So, from where is it being driven? Electronic devices and the applications in which today's information technology finds its gestalt, no longer provide realization in the old sense any more—or so goes our argument—nor do they produce any centrally manageable power capacity as do the engines of thermodynamic apparatus. Rather, electronic application technology channels such power capacity into practical applicability by the many, with many different purposes, interests, intentions. We are therefore not seeing primary physical-energetic motricity in this motricity; rather, we should consider it to be primarily symbolic-energetic motricity. Electronic application technology finds its differentiating driving force in the interest of people who use it to organize their daily chores. Architectonic motorics of the urban folds the fixed point, indispensable for any mechanics, either as an Archimedean point in the without or as a thermodynamic motor, from an earthed and territorially localizable position into the distributed domain of symbolic values. Symbols organize their own effectiveness through distribution, not through reference to some assumed firm ground that was not itself symbolizable in variable ways. Therefore, architectonic motorics of the urban—which as architectonics is chiefly interested in the conditions in which ideas of hold, stability, and order may appear—is not primarily aimed at some construction of "residues" or "stocks" of knowledge, or of the established. Characterized as urban motorics, our architectonics acquires its stability through differentiating these stocks, and in this sense is focused upon a more comprehensive notion of cultivation *(Bildung)*. The more differentiatedly we are able to convert our actions within the urban into sophisticated capabilities to act, the more stable the architectonic motorics.

IX   VALUES

What remains now is the question of the how-functioning of this motorics of the urban. Functioning is always consequential upon the purpose at which it is directed, a fact that poses problems when related to a pre-specific, non-assessed potentiality with which we have characterized our notion of the urban. The prime motive informing the present text lies in the quest for a philosophic attitude towards primary abundance. While such abundance—according to photovoltaics, and electronic energy logistics, in which it is grounded—is a civilization-historical product, to us it is an abundance principle that as such cannot be apperceived following traditional modes of observation. We perceive this abundance principle as pervading our global living conditions with formless potential. Were the phantasma of the perpetual-motion machine not tantamount to asserting independence of the machinal from an external source of energy, one might indeed recognize, in this abundance principle, the principle of perpetual motion. But it sets free a form of potentiality that precedes any valuation and any form, and thereby escapes calculability precisely in the way calculability is being proffered by perpetual-motion machines. The kind of potentiality to be released from primary abundance must be seen as being unconditioned, or in that sense absolute, since being genuinely formless. On behalf of this kind of potentiality, we suggest the language game of virtuality, as will be explained below.

The challenge for philosophic thought of the urban lies perhaps precisely in finding a way out of the perpetual-motion phantasma. Even Lefebvre's ideas of total urbanization remain prisoners of it. He can conceive of his impulse-setting activism in terms of revolutionary dynamics only because his idea of totalization of the urban unhooks the latter from whatever dependency on anything extraneous.[37] In our mind, however, the urban remains kept in constitutive dependency, cultivating culturly urban values as a breeding ground, in the manner of urban agriculture. Design and planning can no longer mutually orient themselves in accordance with some territorial ordering and structuring, as to quantity, scope, and proportion. The usual procedure, of methodically searching for the shortest way as the way of best promise of stability, there loses its immediate bindingness in favour of proceduralizing what passes for valuable. The richness of that which passes for valuable, in this conception lies less in a capacity to conserve the energy-ness of natural-material resources than in conservation of the symbol-ness encapsulated within the cultural-historically grown referential conditions. What passes as valuable for an urban architectonics derives its value from the richness and fertility of the historically cultivated grounds of knowledge and signification, which in their scientific, cultural, artistic, and economical mappings, are prepared within the urban for acquisitive realization by

---

37 The abundance principle, or so it seems, cannot be comprehended within any thinking of totality, nor within thinking of the absolute, precisely because this unconditionedness remains intrinsic to the constitutive dependency on its own cultivation.

individuals. Thereby, we relate Aristotelian poiesis not to nature, but to the culturly urban richness of the thinkable. We did not describe the urban as a machine for transformation and production of power and power capacity, but as symbolic-energetic motorics cultivating a pervasive flow of virtual and pre-specific potential. The architectonic motorics of such urban aims to expand the scope between the two poles, of stabilizing social community, and of differentiating such stabilities.

Such reflecting about the urban, however, would not be motorics worthy of the name, were it not to be supported by a technological basis. That which is apt to sustain such architectonics must not be dependent upon the moral integrity of individual persons, but must be transferred to some extraneity. Let us remember that information-technological appliances today allow technically supported dispositions to be configured on the basis of information, aptly conceived by James Gleick as *how we can know*. We are tempted to call these ways of giving an account "infra-stories", in reference to the etymology of the Greek word ἱστορία (*historía*) for giving an account of things intuited according to one's own perspectival observation and valuation, which in the late-Latin word *storia* turned into a summary term for "knowledge, history, account, tale, story". In the urbanity conceived here, the technically supported dispositions distribute the infra-stories that they incorporate and as which they are constituted, in sequences, series, and as a fabric of integrations, into the populations. Their sustainability needs to be maintained equally in terms of their architecture-turned-symbolic, or short: technical infrastructure, and their narrativity. The established technics today, operating within the substrate of the electrical, supports the integratability of the individually phantasmal into socially lived-out rationality, in the ancient sense of the Latin word *ratio*, for "calculating estimation, establishment of proportionalities". In this sense, infrastructures in the urban are not just constructive girders to narrations, but are in a constituting manner themselves narrative. In the electrical, inasmuch as an abstract element and nutritious substrate of the urban, large stocks of the historically cultivated grounds for knowledge and signification have already vaporized into purely formally treatable symbolicity. Symbols catalyse values into procedurality by distributing them, formally packaged and encapsulated, into a fabric of differential bindingness. Such bindingnesses today exteriorize themselves in the medial networks which are not only electric, but in their narrativity also medial, and issuing from which the information-technological applications are coming up with technically based narrations for organizing our everyday life. They convey to us modes of conduct of a kind unlike any by which preceding generations of technics, in the form of apparatus or machines, convey modes of conduct to us. In order to allow us, nevertheless, to speak of motorics, we argue that this language game be in orthogonal relation to the described and varying generations of technics. This assumption of motoric orthogonality cannot be entirely detached from the assumption of a language game of continuity, because in whatever way, every intermediation, if to be deemed motoric, demands continuity of that which is to be coherent as motorics.

Are we to stick to our motorics language game then, the challenge consists in comprehending this continuity neither as naturalness nor as any other objective reference, but in its mediality. In contrast to geometric-proportional or thermodynamic-transformative mechanics, which presupposes continuity either as nature or as reference, we cannot presume, on behalf of our motorics, that medial continuity might "earth", in a "containing" sense, the constructs it helps to realize. Much rather, a notion of intermediation must belong to such language game of urban architectonic motorics for which the continuity itself that flows from the intermediation's taking place, is an integral part of the motoric activity it describes. Such an idea of intermediation seems only imaginable within an element of abstract "formalness"; it does not take place through value transfer, as in machine power, nor through value transformation, as in the generation of apparatus capacity. Rather, this intermediation happens through proceduralizing values formally and through proffering the formal procedures obtained thus. In this text, we have set out to comprehend the electrical as such an element of abstract "formalness".

X   INVARIANCES

It is not easy to imagine how one might reflect about change and development setting out from the abstract formality as we have sketched it here for electricity without "earthing" or "territorializing" it from the outset. In that abstract space, starting points, planes of reference, and coordinates for orientation are set with no more legitimacy than that of an experiment. If we do not want to "earth" or "territorialize" that space unreflectedly, we are in a space of general modelling without anchor point. As difficult to imagine as this may be, we are familiar with a quite comparable situation in cultural history. Metallurgists managed to distil substances from rock and prove to themselves—and everybody else—that a certain something that was present in the rock could be extracted from it, and brought into any form whatsoever. Furthermore, this shaping, performed upon metals, i.e. upon the "abstract", is reversible. At that time, people began exploring the phenomenon that things, such as may be found and known by them, and for which they have developed an intuitive feeling from daily interaction with them, may be brought into different consistencies, different-looking surfaces, and different shapes. In so doing, metallurgists acted in keeping with a notion of invariances.[38] Via this assumption, we may at least distinguish today's materials sciences from earlier metallurgy. For unlike the idea of homogeneous uniformity of matter, the idea of invariances suggests that materials are never, to our understanding, accessible disconnectedly from the variations in which they materialize. We are of course far from advocating, on behalf of architectonic motorics of the urban, resuscitation of metallurgy and the attending alchemy, scientifically speaking.

Nevertheless, we would propose taking that tradition of considering invariances up again, so as to facilitate proceduralized dealings with the values that are being processed and differentiated within our model of an urban motorics. A formal-analytical toolbox for such metallurgy of urban consistencies may be made out, at least embryonically, already in Lagrange's analytical mechanics. Joseph-Louis Lagrange succeeded Leonhard Euler at the Prussian Academy of Sciences in Berlin, and very likely was impressed with Euler's equations with imaginary numbers, and the concomitant abstraction from an intuitive, figurative view.[39] The core idea of his analytical mechanics of 1788 consisted in the breaking down of all constants of system equations into ever-finer procedures. Lagrange has introduced nothing less than a method for generalizing coordinates. On the strength of it, the assumed constants may, in increasing differentiation, be made treatable on behalf of a multitude of variously differential purposes.[40] The difficulty, in philosophically cogitating the city or the urban, lies, as we have seen before, with

---

38   Gilles Deleuze, Félix Guattari, "Treatise on Nomadology—the War Machine", in: *A Thousand Plateaus. Capitalism and Schizophrenia II*. Trans. Brian Massumi. Minnesota University Press, Minneapolis 1988, pp. 351–423, especially p. 412ff.

39   For a historical account of Leonhard Euler's symbolical strategies for coming to terms with an "algebra of seeing", cf. Wladimir Velminski, *Form, Zahl, Symbol—Leonhard Eulers Strategien der Anschaulichkeit*. Akademie Verlag, Berlin 2009. Velminski sums up what is at issue in Euler's algebra of seeing: "Die Gestalt des Auges wird zum variablen Verfahren, um durch symbolische Operationen experimentelle Praktiken—eine Algebra des Sehens—zu erfinden. Und da Algebra die Kunst ist, aus unbeständigen Gleichungen, Ungleichungen und Identitäten Schlüsse zu ziehen, setzt Euler für Variable Zahlen ein, mit denen er den Vorgang des Sehens fokussiert", p. 229. For a mathematical history of imaginary numbers and their applications, cf. Paul J. Nahin, *An Imaginary Tale: The Story of i [the square root of minus one]*. Princeton University Press, Princeton 2007.

40   Coopersmith elaborates: "Although he doesn't elaborate, this [the "generalized coordinates" VB] is where all the hard physical thinking comes into each new, mechanical problem. One must examine the scenario carefully (whether it be a compound pendulum, a rotating solid body, in fact, any combination of pulleys, levers, inclined planes, and so on) and then use past experience and general physical nous to determine what are the relevant degrees of freedom of the problem. The degrees of freedom are those features that determine and describe all the possible motions of the mechanical system. The generalized coordinates then map out these degrees of freedom. They should form a complete set (not leave any feature undescribed), but they don't have to be the minimum set possible (some amount of over-determination is allowed)." *Energy, the Subtle Concept*, ibid., pp. 136–137. The latter statement regarding relief from granting the minimal set possible is important for our translation into urban motorics, as here one of the troubling questions is how to behave methodically, without relying on the assumption that the "shortest way possible" were the most reliable way, in terms of which method to follow.

the capacity of dealing in an open and open-ended, yet systematic manner with the determinability of initial and final states. It is precisely the determination of such "initial" and "final" states that needs to be proceduralized. And that's what in Lagrange's analytical mechanics is productive for our architectonic motorics. As algebraic mechanics, it neither starts from a known initial or a final state, nor does it treat forces as merely reactive forces but brings them into experimental constellations through the principle of Virtual Work,[41] and is then not compelled to standardize quantitative differences of its measurements in a stochastic-global manner. Lagrange's *Mécanique analytique* deserves great credit for leading mechanics over into the realm of mathematical analysis, and thus for the advent of thermodynamic apparatus. Typically, he writes in his preface: "One will not find figures in this work. The methods that I expound require neither constructions, nor geometrical or mechanical arguments, but only algebraic operations, subject to a regular and uniform course."[42] Had Lagrange, in his reflection about mechanics, not referred to a premise of uniformity, one might indeed say that his principle of algebraic mechanics prepared, in the thinking about technics, a constitutive role for Nietzsche's later notion of *the eternal recurrence*.[43]

In algebraic mechanics, the assumption of invariances constitutes the necessary condition for bringing them, upon the impermanent gestalts of their embodiments, into modelling-induced continuity. Next to Lagrange, it was Leonhard Euler who contributed prominently to the development of the required variational calculus. The reliability of such modelling grows in proportion to the degree of differentiation of the models obtained. The measure of their differentiatedness lies in how many of the variable impermanent gestalts of such invariances may be brought within the model, under one integration, and in stable fashion, into relationships to one another, without the need to reduce

---

41 The principle of Virtual Work goes back to Johann Bernoulli, and is summarized by Coopersmith as follows: "It applies to systems in equilibrium—in other words, where there is no movement between the parts. There has always been a problem with analysing such systems—it is the fact that there is no movement, an indication that there are no forces present, or are all the forces exactly balanced out?" The principle of Virtual Work allows us to remain undecided with regard to this assumption, and in fact provides a frameset for finding out by testing through symbolic operations: "To a system of variously directed forces in equilibrium, an overall virtual displacement is applied. This results in a set of local virtual displacements at the point of application of each internal force. Each force therefore carries out virtual work that is neither positive [...] nor negative [...]." *Energy, the Subtle Concept*, ibid., p. 128 / 9.
42 cited in: Leo Corry, "The Development of the Idea of Proof up to 1900", in: Tim Gowers (ed.), *The Princeton Companion to Mathematics*. Princeton University Press, Princeton, NJ 2008, pp. 129–142.
43 Nietzsche defined his notion of "Will to Power" as that element from which simultaneously emerge the quantity differences between related forces, and the quality that applies to each of the forces respectively, as long as they are within the established relation. Cf. Gilles Deleuze, *Nietzsche und die Philosophie*. EVA, Munich 1976 [orig. 1962], p. 56ff.

their variability. In today's sciences, these mathematics are importantly used in the research of minimal surfaces, e.g. of soap bubbles. We suspect that in analytics related to our architectonic motorics of the urban, it might play an important role. Because variational-calculus modelling, as starting from the assumption of invariances analysable only by means of surface measurements and their behaviour to one another, works comparatistically within the framework of its models and can be evaluated—at least if it were to be portable from scientific mechanics to general motorics of the urban—literally according to the degree of differentiation to which such modelling manages to integrate the fantastic into the rationalized.

XI MEDIALITY

Design and planning based, as their dynamic "foundation", on such architectonic motorics, are consequently unable to obtain mutual orientation from some earthed and territorial, quantity, scope and proportion-related ordering and structuring. The idea of identity had been central to such territorial, geometric ordering. By putting the idea of invariances into the philosophically systematic place of identity, we obtain a new referential frame for orienting design and planning: the criterion of fitness of forms is superseded by cross-compatibility of the values behind the formal gestalts of their appearance. We assume in our motorics such compatibility of values to be, via the mechanics of proceduralizing, accessible to variational-calculatory analysing, comparing and construing. Thus, the gestalts of values assumed as invariances may comparatistically be tested for compatibility. The conspicuous advantage of the swap of roles between identity and invariance, as submitted here, lies in the fact that the number and quantity of such values assumable as invariances are not in any principle limited. On the contrary, such role exchange lets us, in open-ended fashion, differentiate and refine values, and develop products from them. In our view, this is a referential framework for householding with culture; from it we shall see emerging, on behalf of choreostemics, the anchor point apt to give direction to that third gestalt of the technical that we postulated for electronic appliances. Of course, off-hand it might seem obvious to associate the unfamiliar notion of invariance with technical schemata, types, templates, print forms and the like. But that would mean foreshortening the issue and regressing to identity-logical ways of thinking, because the characteristic of these concepts is precisely to make identical things technically reproducible; in this way they are indeed prevented from proceduralizing their cultural value—they particularize it.
As we know from Lagrange, the notion of invariance is borrowed from formal-scientific discourses. There, however, it is *not* representative of reproducible shapes, but is used for treating quantities that are not

conceptually treatable as numeric values—e.g. the numeric values set in applying Lagrange's generalized coordinates are set out of purely symbolical considerations.[44] Their invariability is being assumed so as to enable systematic treatment of measurable gestalts as variations of such modelled invariance—hence also the descriptive term for such processes as "virtual work".

The proposed role swap moves, via the notional couple of identity and invariance, the notions of individual and quantity into interesting proximity. The two notions, that of individuality and that of quantities, share the primary fact that they are referring to some not entirely positivizable "consistency". They thereby postulate a consistency that might neither be taken purely quantitatively, as generalness, nor purely qualitatively, as special expression.

From the relation between individual and quantity, new roles may be foreseen for the mediating language games of values and shapes, by which we shall now be in a position more accurately to describe our motorics. Traditionally, forms are considered to be means of bringing measured values into proportional continuity. From this capacity of the forms, geometric mechanics are evolving. So, what is the appearance of such mediation when seen in the light of algebraic mechanics? It seems to be crucial to extricate ourselves from the language game of mediation-as-"imprinting", or "transmission", and to encourage a language game of mediation as "proceduralization", with some analytical mechanics as starting point. So we shall assume, on behalf of such mediation language game, invariances that we take to be embodiments of consistencies that are never positively, but only differentially determinable. This is not an unusual assumption, and different ideas of approximation logics are normally applied to such analysis. But we are looking for another way, since the idea of approximation remains bound to identity thinking. Thus, let us assume that these "bodies" of invariances as such are not positivizable, they just appear, in the gestalts in which we perceive them, supported by our encoding and decoding habits and expertise. The gestalts of these embodiments become apprehensible only through interpretation, by being brought into continuity with what we know, and what we are thus able to see in them.

Without doubt Henri Lefebvre's thoughts, with the notion of the producibleness of space as a situation generated through creative action, followed the same direction in his theory of total urbanization. But short of carrying formality algebraically-symbolically further and thereby stabilizing structures intended for symbolic encapsulation and integration of released impulses, such dynamics cannot but end in explosive incandescence.

---

44 Re problems surrounding number and quantity, and the intermediary role of measuring the in-between, cf. Augustus de Morgan, *The Connexion of Number and Magnitude: An Attempt to Explain the Fifth Book of Euclid*. Kessinger Publishing, Whitefish MT 2009.

XII   DOUBLE ARTICULATION

If algebraic-symbolic cogitation on formality is to be carried another step further, then it is neither in its symbolically expressed form nor its approximatable-scopeness or, as it is often termed in analytical logic, as an extension. An approach that is pointing to a further degree in abstraction may be found in Louis Hjelmslev's algebraic semiotics. His noetic figure of "double articulation" sought to set the notions of content and expression into mutually constitutive relation.[45] Gilles Deleuze and Félix Guattari then imported Hjelmslev's concept from linguistics and developed, in their text called "Géologie de la morale" (geology of morals),[46] a philosophic language game about formality within the framework of analytic invariance-thinking about values.[47] They generalize, for use in philosophy, Hjelmslev's differential relation of linguistic content and expression, into a relation between "form" and "substance". This justifies my reading here, in which I link their notion of "double articulation" with their remarks about metallurgy,[48] and the

---

45   Louis Hjelmslev, *Prolegomena to a Theory of Language*. Trans. Francis J. Whitfield. University of Wisconsin Press, Madison 1961. Cf. the adaptation of this noetic figure by Deleuze and Guattari: "The first articulation concerns content, the second expression. The distinction between the two articulations is not between forms and substances but between content and expression, expression having just as much substance as content and content just as much form as expression. [...] There is never correspondence or conformity between content and expression, only isomorphism with reciprocal pre-supposition. The distinction between content and expression is always real, in various ways, but it cannot be said that the terms pre-exist their double articulation. It is the double articulation that distributes them according to the line it draws in each stratum; it is what constitutes their real distinction. (On the other hand, there is no real distinction between form and substance, only a mental or modal distinction: since substances are nothing other than formed matters, formless substances are inconceivable, although it is possible in certain instances to conceive of substanceless forms.)" Gilles Deleuze, Félix Guattari, "The Geology of Morals", in: *A Thousand Plateaus. Capitalism and Schizophrenia II*. Trans. Brian Massumi. Minnesota University Press, Minneapolis 1988, pp. 39–74, here p. 44. a

46   *The Geology of Morals*, ibid.

47   They do not themselves call it this, but refer to the aspect of relative invariances within what they call "a function of stratification". Their interest is to think about variance and variability in a way that does not rely on any identity notion. Cf. for example: "Even though it is capable of invariance, expression is just as much a variable as content. Content and expression are two variables of a function of stratification. They not only vary from one stratum to another, but intermingle, and within the same stratum multiply and divide *ad infinitum*. Since every articulation is double, there is not an articulation of content and an articulation of expression—the articulation of content is double in its own right, and constitutes a relative expression within content; the articulation of expression is also double and constitutes a relative content within expression. For this reason, there exist intermediate states between content and expression, expression and content: the levels, equilibriums, and exchanges through which a stratified system passes." *The Geology of Morals*, ibid., p. 44; or more straightforwardly: "A stratum always has a dimension of the expressible or of expression serving as the basis for a relative invariance", *The Geology of Morals*, ibid., p. 43.

48   Which they develop in another chapter of the same volume: Gilles Deleuze, Félix Guattari, "Treatise on Nomadology–the War Machine", in: *A Thousand Plateaus. Capitalism and Schizophrenia II*. Trans. Brian Massumi. Minnesota University Press, Minneapolis 1988, pp. 351–423.

notions of "singularity" and "*haecceitas*" which they use in that context. Broadly speaking, each of these notions belongs to a different line of philosophic discourse. "Singularity", in more formal and measuring theory discourses, means one singled-out relation, one that is applicable to just one field, and not liable to generalizing exportation. "*Hæcceitas*" on the other hand, was introduced, among others, by Duns Scotus for emphasizing the individual features of an object as opposed to its general properties as an element of a class. Singularity, for example, means for Deleuze and Guattari a physical-nomological index, such as the melting point of some specific material, whereas "*hæcceitas*" means instances that are generated through dealing with such indices. In consistency-metallurgy, which they are thus heralding, each technical poiesis always plays in several such varietal lines at the same time.

Thereby, the narrowing that traditionally attends the notions singly can be avoided. On the one hand, individuality can be released from marginalization, seen, under formal aspects, as a shortcoming inflicted upon individuality because there is no scientific discussion of singles. On the other hand, the idea is to pry individuality loose from being mystically-praisefully exalted to perfection, meant to be exclusively experienceable, and in no way expressible.

This, then, is the way of dealing double-articulatingly with the formality issue of culturly urban consistencies. Such consistencies as are being mined, by the Deleuzean-Guattarian noetic figure of the geologist of morals, from culturly urban sediments, and reprocessed for ulterior use, are then, after the erstwhile metallurgists' fashion, pre-specific in a radical sense. This pre-specificity constitutes their virtual status, as a potential value that is only consecutively and procedurally determinable in the course of acting according to that value in the way being determined *in actu*. This sounds rather abstract. But by recalling today's gestalt of the technical as electronic appliances, we find many examples of pre-specific value and their virtual meaningfulness. The architectonic motorics that is at the centre of our investigation here is being driven by way of applying what we may learn to know, surmise, and esteem.

XIII   CODA

The real question now, as posed for a philosophy of the urban as architectonic motorics, is how this energetic flow of undifferentiated potentiality might be encoded and symbolically integrated, so as to allow differentiable capacity to be gained from it that is apt to be developed into ability, proficiency and artifice.

Our initial question was how hedgehog cunning and hare performance might be put in reference to each other without giving precedence to either. The gist of the tale was the competitive situation between the capacities-related principles of distribution, symbols and discontinuity on the one hand, and that of mobility, energy, and continuity on the other. The reconciliatory question, of how to deal with both of them, may now be rephrased to how to virtualize values we consider as actual and binding—a potentially life-saving point to the hare, or at least dignity-saving. For it would have let him forget his physical superiority, and perhaps see through the hedgehog's ruse. On the other hand, and complementarily, there is another question coming up: of how to gain something potentially actual from something virtual. After all, it is neither easy nor evident, sign-technically speaking, to deal with an unfair-contest situation by displaying one's self as specific "information", and thus to turn the fairness issue upside down and regain the upper hand in an uncomfortable situation.

INDEXICAL MARKINGS OF THE TOPICS DISCUSSED

These summary discussion threads relate to the lecture "On the Question of Constructing within the Symbolic", which was presented by Vera Bühlmann at the first Metalithicum Conference, and which forms the basis of this text.

A first discussion thread developed around the topos of networks and their topologies. It was considered to what degree each instance of speaking of networks must always already include a specific dynamic as a constitutive dimension. The discussion went that if such a dynamic were excluded from considering networks, it is possible to speak seemingly much more clearly, in the familiar language of games of structure, frameworks or the like, about the idea of networks, which is still very vague today. These familiar language games were identified as being part of the direct tradition of geometric mappings, however, which form the basis of a classical-mechanical understanding of dynamics, and which therefore are conceived on the premise of initially static conditions. Several views focused on the degree to which descriptions of static conditions can at all be assumed as adequate when dealing with networks;

respectively, how to deal with the fact that such static conditions, when considering networks as networks, in their dynamics, can only be attributed to them—but can never be "observed" objectively, in an uninvolved manner. The situation is similar to that which had led to Lagrange's algebraization of Newtonian mechanics— namely the question of how the stability of dynamic systems at rest can be tested, without knowing in advance whether the apparent resting state can be understood as static or instead, whether a subtle temporary balance of the system's constitutive forces is responsible at that moment. As was discussed, there is a dependency on dynamic systems between the ascribed state descriptions, the interests that prompt an investigation, and the observer's patterns of expectation. If with all due caution, prematurely dismissing the language game of the networks altogether were to be avoided, some interesting questions arise that apply specifically to the established noetic schemes regarding causation, causality, determination, tendency, intentionality or the like.

An additional discussion thread developed around the topos of a genealogy of mediality. As was discussed, mediality has been closely linked with the inquiries into the capacities of objects, animals and people since ancient Greece. Even if the current format of the medium now has very little to do with earlier occult formats—from visions and revelations, inspiration, the muse's kiss, to the obsession with and the idea of the romantic cult of genius in terms of an artist's ingenious, therefore incomparable singularity of thought—mediality and media's influence on an individual's thinking is nevertheless still a topic today, for example in the areas of education and information. It was discussed if, and in what sense, the idea that it might not only be possible to train one's cognitive abilities within a naturally given disposition of capacities, but that it might also be possible to learn how to expand, compose, refine these dispositions themselves, poses a challenge to the values of enlightenment, for example in societal institutions of education in their established forms. In sometimes very heated discussions, differing language games were considered for this re-emerging philosophical dimension of capacities and capabilities, and especially also the role of technology therein; it was suggested for example to consider an "unsettled capacity for thought", or an "intervention of novel disponibilities of cognitive capacities".

A third discussion thread developed around the topos of a concept of continuity. In particular, the discussion centred on how to find a way out of the discursive calcifications that arise from schemas of linearity, circularity or an absolute break within historical thought. At stake are two incompatible postulates: on the one hand the assumption that the past and future can be seamlessly reconstructed, and on the other, the assumption of a possible innovation, caesura, a starting era, or the assumption of modernity. The discussion as to whether and to what degree the concept of the symbol can be rendered productive in order to emerge from a dialectic view of continuity and discreetness was relatively controversial. Regarded from an application oriented or operational perspective, the concept of symbol has always been involved, as Bühlmann argued, allowing the conditions of continuity to be created. *Symbolon* (symbol) generally means "that which is thrown together". According to this view of the conceptual history of the symbol, this concept has always referred to a situation that is constituted by an initially unsettled and dynamic event. A *symbolon*—in the Greek sense of a guest label, or carrier of another contractual agreement, only functions by investing a tear or a discontinuity with meaning and by indexing it for verifiable agreement. In this ancient concept of symbol, the discontinuity itself can neither be reduced to a break (which it is) or as a merging (which it is as well), but instead, it allows a handling method, an operability that simultaneously produces that which it wishes to secure—without its being assumed to exist independently of the implementation of this handling method. The discussions focused in particular on how these specific operational capabilities of symbols could methodologically be integrated into a broader values debate. Opinions were especially divided on these points. In summary, it can be argued that if a sign-interpretive rather than an operational perspective is selected, the genealogy of the concept of symbol is cast in a light that appears to threaten hermeneutic-contemplative semiotic-theoretical thinking at its core: if the symbolized relations of reference are granted a positivity, as a universal value for example, as an archetype, a thought-form (and not a figure of thought) or the like, they indeed arrest the process by which the meaning of signs emerges as open, yet analytically accessible, semiosis. A consideration of the operability of symbols, however, as Bühlmann counters, is like a semiological perspective as well, directed against the assumption of a positivity of referential relations. Yet the application-oriented, operational view of symbols attempts to align itself with the "effects" which, as a result of an interpretive signs practice, inevitably keep arising, appearing, and challenging.

A fourth discussion thread developed around the topos of "projective construction", or "acting by designing". The dynamics, to which the *symbolon* ("that which is thrown together") refers, is one of throwing. From an etymological perspective, the Latin term "to throw", *jacere*, constitutes the central word stem for the concept of "object", of "subject" and of "project". In this sense, for every composition or design, a projective or an investing, credit-giving plan would have to be considered constituent. The concept of symbol in Bühlmann's presentation suggests dealing with this by way of an active indexing or placement of markings. Certainly the liability of this kind of concept of symbol can only result through the ceasing or missing acceptance of what was meant. Projective design is formative, in this case, because it must involve the acceptance conditions in its design. These discussions are also controversial. In particular, a more administrative perspective following Heidegger's postulate of an "axiomatization of composition / design" was cited against the operational concept of the symbol, and its pointing in the direction of design thinking conceived as a way of economic cultivation.

IV THAT CENTRE-POINT THING — THE THEORY MODEL IN MODEL THEORY
KLAUS WASSERMANN

I UP AND IN 159 — II THE DEMISE OF THE MIDDLE (LONG LIVE THE MIDDLE) 161 — III MODEL 169 — IV DE-CENTREMENT OF MODELS 171 — V THE PRACTICE OF THE DE-CENTRED MODEL 175 — VI THE NON-ASCERTAINABLE LOCUS 181

KLAUS WASSERMANN studied biology (the psychosomatics of stress, the complexity of social systems, brain research, methods of statistical modelling) and philosophy (philosophy of language, epistemology) at the University of Bayreuth. He held a position as researcher and lecturer at the mathematical-biological unit at the Chair for Psychology at the University of Zurich. From 1994 he worked as a software architect and product manager in the field of machine-based reasoning and its practical implementation. Since 2009 he has been senior researcher at the Chair for Computer Aided Architectural Design CAAD at the Swiss Federal Institute of Technology (ETH) Zurich.

Our reflections address the question about the conditions of the possibility for a machine-based episteme. Wrapped in this first question comes another, only apparently technical in nature: how can we speak about the whole of the field into which we are striding? For we shall find there not only the complete clutter of what today goes by the name of information but also all that is vaguely related to it. The initial question itself is in turn embedded, Russian doll-like, in a large historic movement, which, too, becomes consequently part of the setup and which we might—in neat demarcation from decentralization and deterritorialization—call "de-centrement", attended perhaps by a hunch that this abstract notion might turn out structurally effective, and helpful in disentangling our thinking from geometrical shackles and mesocosmical gaggings. That historic turn towards ever-increasing decentrement challenges any foundation, rules, structures, procedures and patterns that served so far for apprehending the world, in particular geometry and corporeality. Our very way of life today does not merely compel us to conceive of our ontogenesis in terms of autogenesis, and hence of ontology as autology. We furthermore face the need of specifically calibrating anew—or for the first time?—the role of the model itself and thereby to arrive at a philosophical meaning of information. This implies theoretical movement, and consequently the threat of entrapment in the notorious onto-epistemological vicious circle—a circle that is in no way a mere delusion, as we shall demonstrate. We may well find it must indeed effectively be overcome prior to any answer at all to our inquiry into what conditions determine the possibility of a machine-based episteme.

I   UP AND IN

Secularization is, or so it is said, the profanizing of hallowed structures, with sanctity being dragged into and upon the world. Hallowed, in this game, is tantamount to otherworldly regarding both origin and effect. Hence, the term relates definitively to, and declares, the noncorporeal. Any secularization therefore seems to be an assertion of some territorialization, in consequence of which hallowedness, the unattainable immaterial, finds itself landed right in the middle of corporeality, indeed of physicality with all its fluids and mechanical whirls.
Ultimately, the dialectics of this "here and above" were and are leading time and again up against deep cultural ditches, some of which are as yet unconquered and still operational. No wonder, one might say, since arguments are rarely effective in the face of hallowedness, if indeed appropriate. The transition from the over-sphere, scale-free and eternal, to the sphere of the corporeal was at all times fraught with problems, not just in one direction, as for Moses and Muhammad, but at least since Plato in the other direction as well. It is striking that to the three of them, Moses, Muhammad and Plato, transition was functionally a one-way street whose constraint was to be overcome but through death or some other bodily discontinuance. Whereas our own plea is more in favour of bidirectional bilaterality. Hence we ask: how do we come by ideas; what is an idea anyway; where are its premises rooted; and, first of all, what does it take for the body, this array of matter, to arrive at an idea, and a communicable one at that?

Dear friends, dear colleagues, the incomprehensible, that to which we are unable to connect any idea, quite obviously lies in wait at every corner, which is why we not only cannot get rid of philosophy, but are in real need of it at each of those corners. Strong arguments exist for a revival, at least a temporary one, of philosophy, especially philosophy of the episteme. One of the strongest comes from that technology we are calling—still somewhat clumsily, certainly also a tad epoch-chauvinistically and definitely not quite appropriately—"information technology". My interest goes, both in general and in this presentation as well, to the relationship between machine and idea. Or, more specifically within the present context, to the conditions determining the possibility of a machine-based episteme. Off the bat, placing machine and episteme so closely together may sound surprising, it may indeed dent this or that taboo. Much more conventional would be the question about calculability, or about the limits of calculability applicable to this or that context. And the famous question about verification of that which we call knowledge would not be far away, either. Nor would the argument about the difference between digital and analogue.

Machines pass as being determined. So do computers. This is why a computer would never be capable of doing, nay learning something "on its own": because—or so the assertion goes—it was capable only of what capabilities were programmed into it. Since recently, even the most complex construct in the known universe, the human brain, has now made it to the status of determinacy, at least in the minds of the joiners of a certain trend of radical naturalizing. Even as most of this discourse's arguments can be traced back to misinterpretations, or categorical errors, there yet remains the philosophical problem that things might all the same be that way. How can, within determinism, or a world just consisting of mechanisms as well, more be known in the end than in the beginning, when knowledge must be a true and justified belief,[1] while substance-extending deductions cannot be reliable?

In my mind, probing into the premises that might sustain a machine-based episteme may well be the only way of carrying those discourses one or more steps further, discourses that are probably at present growing livelier because the technology in today's machines is so deeply, widely, and intentionally related to information. Therefore, when asking about the nature of information, our view must extend over and beyond the shapely horizon of our electronic soup tureen—both ways, of course, out and in. Along with Guattari it may be surmised that the new machinal

---

[1] Gettier first argued that this conceptualization of knowledge is flawed. His arguments are famously known as "Gettier problems". One of the most convincing answers has been given by Isabelle Peschard, which is based on the idea of language games. Gettier E.L., 1963, "Is Justified True Belief Knowledge?" *Analysis* 23: 121–123. Peschard I., *Reality Without Representation. The theory of enaction and its epistemological legitimacy.* Ecole Polytechnique, Paris 2004.

might perhaps, too, afford a serious chance of better understanding the episteme itself, i.e. at once the capacity and incapacity of knowing inasmuch as a figure of the human setup.

Nevertheless, I shall not be attempting some epistemology that might be termed "philosophical". The reflections below about the conditions of knowledge are upstream of philosophy, a sort of proto-philosophy as Deleuze called it,[2] but—due to their technical components—at once also downstream. Yet, nor shall I hand epistemology over to the sciences or the IT engineers, as was originally suggested by Quine[3] and is ending up these days in a material fundamentalism of sorts as well as often absurd neuro-totalitarian discussions. In that vein, there is serious cause for fear that someone will turn up soon with some neuro-architecture or other—an architecture in which the materiality of our neurological apparatus is seen as a limiting, nay determinant condition.

II THE DEMISE OF THE MIDDLE (LONG LIVE THE MIDDLE)

As the body of living creatures, ours is rather material. Put this way, that sounds trivial, but the very persistence of the subject indicates that things turning around this body and this corporeality might not to be that simple. All cultures at all times have unrelentingly addressed corporeality under multitudinous aspects, be it in more general social discourses, be it in more academic ones. Since that body lives in a flat three-dimensional space, thousands of years ago we already happened upon the notion that this body in its materiality was an essential part of the very person himself or herself. Ever since, the body has time and again been perceived as the quintessential carrier of a person's nature, or at least as one of its most important co-carriers. This materialistic delusion, contrived into a virus by Marx and Hegel, is widespread even today. I shall ignore here all the ostensibly secular discourses about cult and body, prey and exploitation, gender and embodiment, that either respond to this delusion or evoke it, but that in any event go hardly if at all beyond it. Anyhow, this middle-is-matter story begins with Archimedes—at least in my narrative foreshortening here, possibly less committed to historical correctness than to metaphor.

At that particular time, Archimedes very clearly marked those circles as being *his* circles. After all, to him it was not so much abstract sketching that brought those circles about in the sand, than concrete performance. Hence, he linked the geometric figures to his body and, inversely, the execution itself, the action, to geometry. Seen that way, the circle around the body and the body within the circle have, in Archimedes'

---

2   Deleuze G., *Was ist Philosophie?* Suhrkamp, Frankfurt 2003.
3   Quine W.V.O., "Epistemology Naturalized", in: Kornblith H. (ed.), *Naturalizing Epistemology*. MIT Press, Cambridge (Mass.) 1994, pp. 15–29.

case, already come quite close. Such must also have been that unfortunate soldier's perception when, failing to make a difference between the circles, the body, and the person, he resented Archimedes' intervention as insubordination—and simply stabbed him.

Howsoever, geometry, having thus become anthropogenetic, has ever since determind, either within the scale of the human body or by means of it—well, pretty much everything, at least in our culture. As the rather lofty constructs of ideas or gods very soon were made to yield to the transcendental universalist attraction of line and circle, astonishingly even rather down-to-earth law concepts too—as in Hobbes or in Kant—were directly founded geometrically.[4] No real wonder, either, given that the very logic itself, from its beginnings, harked straight to anthropogenetic geometry. Ever since, the two have formed a most suggestive cast. In full consistency, Ptolemy cast the notion of the Earth as the centre of the whole universe—an act beyond verification by simple reference to empiric data. And an act remarkable in being one of the first to subdue interpretation of nature to that anthropo-chauvinism inherited from various religious myths.

Leaving Ancient Roman culture aside, that first cultural culmination of modularity we have met since the late Middle Ages, an uninterrupted chain of developments and transitions that were often described as "decentralization"—i.e. "taking out from the middle"—of which Copernicus is doubtlessly emblematic. Yet, this perception falls short. For in the meantime, it has become impossible to speak of "decentralization" or "deterritorialization",[5] so my suggestion. Because regarding both notions, an imaginary scale subsists, on which the centre and its periphery are placed in harmonious opposition, on which dwell the territory and its border, the individual and the universal, the corporeal and the divine. Copernicus moved the middle away from our globe all right, to the sun—the idea of the middle itself, however, remained unrestrictedly in force. Not unlike in the language game of decentralization. Here, processes, functions or setups migrate away from the apostrophized middle, they may even largely dissociate, but the middle as the basic concept remains unalteredly in force. Nor is deterritorialization directed at doing away with the notion of a middle; it just seeks to do away with the notion of fixation, e.g. by impeding the setting-up of borders. The result is individualization, a bursting of the controlled middle into multitudinous centres, as in dust or flows.

To my mind, the movement we have really been observing for the last thousand years or so is massively more radical. To my mind, it is about no less than the gradual rescinding of that very scale itself down to its complete demise, the scale that ultimately was alone conferring a sense upon

---

4   Kant I., 1797, *Die Metaphysik der Sitten, Einleitung in die Rechtslehre*. §E. In: *Werke in zwölf Bänden*, vol. 8, Frankfurt am Main 1977, pp. 336–351.
5   Deleuze G. and Guattari F., *A Thousand Plateaus*. Continuum, New York 2004 [1980].

those notions and dialectic pairs. The disappearance of this very special ideate scale—the scale that enables or forces us to speak and think in terms of centres, middles and middle points, of the absolute and the universal— this disappearance I will for the moment call de-centrement.
Metaphysically, de-centrement means incapacitation of any assertion of the possibility of absolute centres. Somewhat hypothetically, the middle, particularly the middle in absolute or transcendental terms has, for roughly the past one thousand years, been on its way out. And in the long run, perhaps the whole compound of centralized thinking may be dissolving.
Against this backdrop, it can no longer come as a surprise that not until believing in the centre had been given up did the totally different precondition for that amorphous environment come about that today is termed mediality. Our life form needed first to de-centre itself sufficiently before we became able to perceive mediality.
De-centring phenomena are essentially predicated upon prior fundamental dissolution of absolute identifiability, whereby this dissolution itself may be described as de-centrement. Things remain absolute, i.e. centralistic, logically identifiable and recognizable, only as long as alternatives are absent. As living environments begin overlapping, and possible living modes begin splitting up and intertwining—in short, as life thickens—we are rapidly being faced with the necessity of getting things fastened or, in other words, constructed. The thing-as-such then disappears, even in relation to its aprioric possibility. Whereupon, however, procedures and models are wanted, and where upstream of model and procedure there is pure, or essential, nothingness. As long as the singular *fiat lux* event was situated at the beginning of time, i.e. rather far back, no procedure nor model was needed. Everything progressed in an orderly fashion, planetary physics was indeed a workable proposition for large parts of social circumstance. Everything was set and unclosed, substantiation less than superfluous. Today, however, it is not so much the extant light, along with its origin, that must be seen as the decisive component of existential effects, than the *"fiat"*, the "let there be" part.
Thus we record: densifications, superimpositions and accretions dissolve the possibility of a priori identity; likewise, notions and relationships fray at their rims even more than they are doing in any case from their middle. This movement sets in—as always, one is tempted to say—with the cities and in the cities, be it Rome, be it the spate of city foundations in the 12th century,[6] be it in Manhattan. In London, in the Netherlands, and in some cosmopolitically densened mathematical minds a modern concept of probability invents itself, within the short time span between 1685 and 1693,[7] in reaction, as it were, to the then emerging new Lebensformen.

6 Benevolo L., *Storia della città*. Editori Laterza, Rome 1975.
7 Hacking I., *The Emergence of Probability: A Philosophical Study of Early Ideas about Probability, Induction and Statistical Inference.* Cambridge University Press, Cambridge 1975.

Of course, probability in its presently prevailing acceptation as "statistics" still labours massively under antiquated centrement symptoms—just think of Gauss, his Normal distribution and its ensuing mean value, which pervades our everyday life in normative obsession.

When speaking of densification and superimposition today, interlinking and networking will promptly be mentioned. The net and netting are ubiquitous, reflecting obviously certain aspects of today's Lebensformen. The notions of net and netting or networking, however, fit our purpose here only in very limited fashion. Appropriate use of the terms requires in each case a highly technical close description—in other words, a massive notional centring. The qualities of different nets of different topologies are too divergent. They reach from totally passive and stupidly logistic as far as autonomous, active and associative, indeed from mineral-like to almost-living. Some network theory may be of technical or political interest, but will hardly, if at all, go beyond factuality.[8] Conversely, networks may be read as manifold tendencies of de-centrement. The same goes for the dialectic between module and figure, or for the (projectional) duality between digitization with its modernistic dust of noughts and ones, and the "assertion" that the world "be" analogous.

In my present view, and philosophically speaking, de-centrement is about the impossibility of the universal, and the negation of identity. As such, the term carries an important historico-political dimension as well that, however, will not be pursued here. Let me just mention that a big part of all social conflicts of the late Middle Ages, along with a large part of the novel positive insights of the so-called natural sciences, or technosciences as they are called today, rally painlessly around that principle. From the string of more remarkable events there are links, in no particular order, to Descartes, Copernicus, Leibniz (of whom more later), the failure of Louis XIV and other such sun kings, the French revolution and the invention of population, thermodynamics, Napoleon, Boltzmann, Bohr, and Web 2.0, as well as to the presently prevalent idol-worshipping, and notorious collective intelligence. Thus de-centrement, as both a metaphysical and trans-European programme, were an essential part of all declarative fields of all sub-cultures, even there where modern dictatorships arose, who in the end proved to be unable to resist this tendency of de-centrement, too.

As you may have sensed, the idea of de-centrement, in all its slightly unhinged novelty, produces, as does every idea, a plane of consistency, provided it be just used sufficiently often and densely. Herein speech compares quite closely to painting or sculpture. Indeed, ideas get denser as they get closer, one might speak of densitappropinquation ("sie dichten sich heran", as used in a Heideggerian way). I am not about to bring

---

8   Barabasi A.L., *Linked: How Everything Is Connected to Everything Else and What It Means*. Plume, New York 2003.

rhetorical-cognitive subliminalia into play, for in opposition to those the problem at hand is of great importance. The first question is of how our ideas gain their reference. It is, however, not to be treated independently, as the philosopher of science van Fraassen has now been trying for years. Actually, we are facing a whole field of problems, part of which is the question of how to use the notion of de-centrement sensibly, seeing that any notion seems indeed to perform the very opposite of de-centrement. A notion, when it works, generally sets things straight, pegs out a claim, centralizes things, and provides them and itself with a mutually generated stable reference. It appears to be that this unavoidably holds true for every notion. But it is not.

Before getting (for a few moments) to the more technical part of my observations, I should like to chart that problematic field briefly. Thus there is, on the one side, as we have seen, de-centrement as a tendency. If we set any store by it, we must perforce discard offhand any statically universal solutions. Without universals there are no more Archimedean anchor points, either for the theory of knowledge or for its practice. In so far as procedures replace the fixed stars that shine in our geometrically formatted parietal lobes, we must resort to modelling, in which, moreover, de-centrement precludes any reference to any fixed locus—also in abstract sense—to any scale, or to any fixed unit of measurement. The loss of the absolute and the universal that de-centrement so clearly brought to light, furthermore requires that we radically volatilize our modelling, i.e. comprehend it from the beginning as multitude and diversity, both when executing our modelling and when referring to their results. In consequence, the result of such modelling cannot not be singular any more.

Phrased differently, we have to get down to regulating the rules. To follow one rule may still be compatible with our Lebensform, but we have all been juggling—and not just since yesterday—with masses of rules regarding which, by the way, it is not in the least clear that they should be addressable in the first place.

Furthermore, we are configuring our societal rules today rather differently from just a few years ago, and we do it perhaps contingently but far from inconsistently, for what reasons whatsoever. One of the less gratifying examples may be found in the accelerating erosion of the democratic Lebensform, as could be observed over the last fifteen years. Regulatory density thickens, as the saying goes. The urge to leave a footprint is part of being human, to a degree. If this person is a parliamentarian, however, leaving a footprint means producing laws. But the idea of a rule for regulating the rules has well lost its horror for us in the "new maze" at least as long as recursion is not progressing further. Pretty soon we might be offensively cultivating self-reference and its paradoxes.

For modelling itself—which compared to Lebensform seems ever to be a rather technical affair, which goes for the humanities too, of course—regulated regulation of rules means that the simple assertion that one has a model will not do any longer, however convincing the tone. Nor will pointing out that everything effectuated "be" based upon models and "be" generating models, while disregarding the before and the after, the conditions of instantation and ethical dimensions. That deficit is widespread not only in the natural sciences. The same embedment-related blind spot at times sprouts curious blooms in philosophy of science as well, as in the attempted differentiation between theory and model, which up to now has always been caught in a loop.[9]

Thus it is not just the notorious side conditions that now become apparent, that *figure de pensée* bequeathed to us by higher differential calculus. No, it is the inherent instantiational conditions that are gaining importance in present and future discourses. Simple ascertainment of thinking, doubt and corporeal being won't do for us any more. It is also why these days one has begun asking about the preconditions of sainthood.

Such instantiational conditions of a rule are not just no part of the rule. They generally lie so far outside of the focusing intention that Kant, e.g. for the case of physics, declared causality to be a priori to pure reason, i.e. a transcendental ideal. Personally, I should like to stay clear of this kind of construction, as such absolutization is tantamount to centralization or centricity which, as we have now learned, ultimately leads to serious problems. As for the regulations of a rule, I would rather say they are situated in a further dimension. Just think of Abbott's Flatlanders,[10] and we might say that the instantiational conditions of a rule are placed orthonormally to the rule itself. Rules are not only regulated, they are orthoregulated. Orthoregulation does not refer to the cross-affecting couple given by the world and the rule, but only to the rule itself. Every rule, all modelling, refers, by way of its mode of construction, to premises, to a priori conditions. The same goes for orthorules. From the model's viewpoint, and sometimes also from the modelling person's, these forerunning settings are almost inevitably quasi-transcendental in nature and therefore all too often remain ignored.

Just consider the perception of causality, all the ifs and thens, the ands and ors, the logical impenetrability of objects. For the time being, we are—in a first degree sense—still unable to walk through walls. And then consider, in contrast to that logicalized impenetrability, the idea of information which is entirely incompatible with that of causality. In the one case, we have identifiability—we are speaking of the visible,

9   Martin Weisberg writes for instance: "Many standard philosophical accounts of scientific practice fail to distinguish between modeling and other types of theory construction. (...) I argue that modeling can be distinguished from other forms of theorizing by the procedures modelers use to represent and to study real-world phenomena: indirect representation and analysis." Of course, such a muddle does not help very much. Weisberg M., 2007, "Who is a Modeler?" *British Journal for Philosophy of Science* 58: 207–233.
10 Abbot E.A., *Flatland*. Echo Library, Teddington 2007 [1884].

and as such concrete, cause—in the other case, we have not. Causality as a notion is centred, information radically de-centred while yet far from coincident with the incoherence of white noise. The nature of information has very little in common with Wiener's and Shannon's reductionist concept of information as defence against white noise. Ever since Kant and Laplace, we have been using the two concepts, not only in parallel to each other, but most often copiously intermingled, with frictions resulting, in biology and medicine as well as in law. At the ascertaining of logic, we have found ourselves, since Kant and Laplace, in an apriorizing twilight zone. In philosophy there rages once more an increasingly intensive discussion about the possibility and the mode of basing knowledge, thereby manifesting almost seismographically an insecurity with which we must yet learn to live. If we do want to come clear about episteme, or such things as calculability and models, computers and simulations, then, I believe, we have somehow to get away from that epistemic twilight. Ignoring the problem will not do. Evolutionary pressure, primed financially through tangible benefits to simulation literati, is too strong for that. However information were to be apprehended, de-centrement would have to serve as leitmotiv.

But let us return to the instantiational conditions, for they are a tough problem. How is the possibility of knowledge to be substantiated or even verified, without resorting to further methods when instantiating such substantiation? This difficulty reflects primarily a failure of the analytical varieties of epistemology which seems never quite conscious of its own speaking, in astonishing opposition to Wittgenstein's findings about our embedding in language if not actually negating them.[11]

The strongest prerequisites of modelling, every kind of modelling, are signs, i.e. the semiotic-conventionalistic relationships along with their mediality. It is, for example, totally impossible to generate a model without a sign system which, indisputably, is convention-based and restricts and modifies what is sayable, or expressible, by the model. Besides, there are weaker, more crafts-like determinants of modelling, such as the mode of logic applied, or the method. Prior to modelling, or to evaluating a model, it wants resolving whether one is indeed prepared to go for a binary model of truth values, or rather for the concept of partial truth values; whether to concede states to the constructed world rather than imagining it as being built of streams. While such fundamental decisions are more of a handicraft nature, as we have seen—there are alternatives, after all—they are nevertheless fundamental in the sense that there is no more catching-up with them in the further epistemic course. The choice of logic for a constructive try can be likened to a secular a priori. It is obvious that the practical consequences of that sort

---

[11] cf. Beckermann A., *Analytic Introduction to the Philosophy of Mind*. de Gruyter, Berlin 2008. Davidson D., *Truth and Predication*. Harvard University Press, Cambridge (Mass.) 2005.

of decision are not easy to assess. The latter determines e.g. in what way, or whether at all, an observer in the classical sense might be admitted. Interestingly, for the relationship between model and sign, the opposite applies as well—how could it be different... A sign or an idea without a model is but an arbitrary name, not even a symbol. Flusser postulates the jettisoning of names as a method,[12] all the while unknowing what the sound of some name might one day symbolize or indeed mean.

The concept of de-centrement as a tool thus enables radical questioning about orthoregulations which, by the way, always carry a metaphysical taste. Put differently, the question regarding orthoregulations is quite in the line of that great tradition of continued de-centrements, and as such is itself not a phenomenon that might be called "new".

It is convenient here to bring the machine-based episteme to the fore again. Commonly, that field is still called artificial intelligence, nay artificial life, or at least machine learning, depending upon the respective author's vanity. Yet, it can easily be demonstrated that no learning can happen if the actor, be it human, parrot, dog, cat, or, well, machine, is not capable of modelling on its own, hence of autonomously developing a theory that might function as orthoregulation for his, her or its modelling. Short of such personal modelling, learning degenerates into parameterization. Which of course is a totally different pair of shoes. Not without reason are we not—yet—saying that we go to school to be parameterized. Here, by the by, there seems also to emerge the only workable definition of the meaning of "theory": theories embody, as language games, the orthoregulation of modelling.

The machine-based episteme, inasmuch as a concept, ought not to be restrained to the review of simple automata. If we perceive, along with Félix Guattari, the machine as the result of subjective history,[13] the thin partition between machine and mechanism disappears. Biochemical mechanisms and their informatization then live in the same field. The genesis of a capability for episteme is not a problem located within the body nor within any single body. Then, we may leave behind us the dualism between information and matter, mind and body, or in quantum-mechanical terms, between wave and particle. Nay, we *must* relinquish those dualisms, if we presume to understand the nature of generalized episteme.

If a statement is to be made about where calculability ends and convention begins, where logic ends and the idea begins, where the mineral mutates into bits and finally into mind, then episteme must be de-centred. The totalizing human reference must be expunged from it. Otherwise, or so it must be reasonably thought, we shall not be

[12] Flusser V., *Dinge und Undinge: Phänomenologische Skizzen*. Hanser, Munich 1993.
[13] Guattari F., "Über Maschinen", in: Schmidgen H. (ed.) *Ästhetik und Maschinismus. Texte zu und von Felix Guattari*. Merve, Berlin 1995.

able either to see or to apprehend the specific dynamics between centrifugal automatization and centripetal automat-becoming. The question cannot be any longer how human knowledge is to be verified. In Eugene Thacker's words, knowledge has for a long time now become a-human.[14] We must take care that the a-human is not suddenly tipping into a highly sophisticated non-human.

III MODEL

So let us now turn to the model. Models take a precarious place between ontology and epistemology, and thereby an important position in our relation to the world, since seen both physically and logically, it is only through models that we are in touch with the world. In logic, the world structure can manifest itself only based on this twin articulation. "World" here does not just mean "totality of all facts", hence "totality of all of which we can speak", but includes beyond that also everything that we (or "nature"), prompted by some manifestation, may shape and arrange. Models precede every empiricism and every practicable logic.[15]

The micrological reason for this lies in the necessity of apperception or ascertainment. We take something for certain, because at this point proving must come to an end, as Wittgenstein put it. But instead of ad-certain-ment (Wahr-Nehmung, literally taking something as representative of truth), one might also say certi-fication (Wahr-Stellung, literally setting something down as representative of truth), for apperception is by no means an exclusively passive event overtaking us, as Leibniz already demonstrated.[16] It is active capturing, deriving from the Latin *per-cipere*. In French, the root of this word has links to tax collection. Putting it less prosaically, apperception becomes measurement. Measurements always point back to the corporeal and its consummation, which is beyond further verification, but which due to its material qualities nevertheless performs a filtering action upon every causal or informational impulse. A filter performs a specific transformation. Therefore, filters are equivalent to specifically applied models. Modelling is inevitably imprinted upon apperception.

In our context, models are seen as proceduralized, hence pre-specific as well as specifying, constructs for that which long ago, prior to Putnam's

---

14 Thacker E., *The Exploit: A Theory of Networks*. University of Minnesota Press, Minneapolis 2007.
15 Johnston C. 2009, "Tractarian objects and logical categories", *Synthese* 167: 145–161.
16 cf. Simmons A. 2001, "Changing the Cartesian Mind: Leibniz on Sensation, Representation and Consciousness", *Phil. Review*, Vol. 110(1): 31–75. The Leibnizian position on perception has been adopted meanwhile also in cognitive sciences, e.g. Thomas N.J.T. 1999, "Are Theories of Imagery Theories of Imagination? An Active Perception Approach to Conscious Mental Content", *Cognitive Science* 23(2): 207–245.

Twin Earth,[17] was called "representation", i.e. referentiality to the world. Yet, models do not represent, they cannot count as extensions of the world.

Hence, the views and representations we develop about the world are in no way, and contrary to the opinions held by some neo-Kantians in their language-philosophical fits, indexical signs. If they were, it would amount to a *petitio principii* or a circular argument, since someone or something would have to build a corresponding index, and this would require rules, models etc. Therefore, models are not indices to the world. Rather, model and world are decisively uncoupled. The only things that for us maintain a link to the world are models, together with corresponding experiments and objections against the backdrop of a will for anticipation.

Interestingly, in this misunderstanding, the Kantian conception of Anschauung (sense-perception, intuition) and logical empiricism look very much alike. The positivistic and modernistic philosophy has considered the simple sensorial data, or the translation of such data by means of measuring instruments, as sufficient verification for knowledge. Any constructive element was denied, falling back even behind Kant. Such misunderstanding is still abundant; for instance, it would be wrong to state that the behaviour of the relations within a model can be conceived as a template of the facts or "surrounds", which should be characterized by the model. Such concept either is reducing or even denying the particular structure of anticipation and its conditionality rooted in conventions, or it is an empty tautological utterance, shifting the burden to the invoked "template", which in turn requires nothing else than a model.

Furthermore, all instances of what we call intersubjectivity depend upon models, respectively a specific culture of modelling, i.e. upon the orthoregulations of our modelling. Now, taking intersubjectivity to be an effect of the notorious mirror neurons denotes definitely massive helplessness vis-à-vis the immaterial. In reality, it is: I am modelling, therefore I am. Descartes' thinking or doubting is just an opaquely phrased special case of modelling. The reason for that is simple: we are fitted with a material body. Since matter serves as the basis for abstract informational processes, or put differently, since our thinking takes place in a body, we must explain the gap between body and information in a manner that refers to rules. Prior to thinking or doubting, dear René therefore is modelling, and be it just with his auricle that already dramatically transforms the sound waves. Remarkably, rules are never just private, even for brainless forms of life, rule-following is possible only and exclusively in some kind of community.

17   Putnam H., "The meaning of 'meaning'", in: *Philosophical Papers, Vol. 2: Mind, Language and Reality*. Cambridge University Press, Cambridge 1985 [1975].

IV DE-CENTREMENT OF MODELS

If we are to avoid ending up in constructive contingencies, modelling must not refer to absolute conceptual centres. Neither direct apperception nor simple sensorial data suffice for verifying knowledge or the ability of knowing. Amazingly, precisely such statically absolute references, such Archimedean fulcra of mind must be abandoned, short of exposing ourselves to an universal relativism of values. In other words, the bottomless and de-centred modelling is our civic duty for the very reason that otherwise we would be winding up in a theocracy, and without a sensible argument against it, at that. Removing the possibility of absolute anchor points, un-Archemeding of modelling actually means de-centring of modelling, i.e. de-centrement of our referential system as it applies to the world.

Any modelling has two sides, time-wise—a before and an after. So far, we have only addressed the before. The after is about the dimension of the accuracy of models, one might think. Getting the de-centrement of the model further, however, requires shedding that talk of model accuracy.

Because the question of whether models are accurate does, for at least three reasons, not arise; nay, it cannot even arise.

In the first place, prior to any modelling, there must be a purpose; model and purpose are intimately intertwined. Where there is a model, there is a purpose, and vice versa. Purposes do not require intentions, they could be embodied almost as immanence into the competitive dynamics of populations, as happens in natural evolution. Purposes may be unrelated to anything crisply identifiable or measurable, as they will be when preference is given to social cohesion. Purposeless contexts must be very rare. They may most likely be found where something is seething and novelty is bursting forth from it in the form of transcendental meaning. Deleuze describes this neatly in *Logique du sens (The Logic of Sense)*.[18] Yet, in this seething there is precisely no more modelling, since there is just seething. In all other cases, the purpose determines the model. Once the purpose is set, the rest can optimally be solved by an automaton within an obligatory and particularly chosen attitude towards risk. We do not want to deny here that fixing a purpose could be an easy thing, for there might be conflicting ones, where the conflict cannot be resolved on empirical grounds. Yet, if it is possible, or if it has been done as a matter of fact, then an automaton can perform the modelling, and it can do so often much better than humans could do.

Secondly, asking for the accuracy of a model is not on, because all modelling is seriously restrained by the quantity of available observations. Generally, in most cases there are too few secure observations for models with

---

[18] Deleuze G., *The Logic of Sense*. Columbia University Press, New York 1990.

workable, i.e. reasonably secure classifications. Every model is more or less wrong, rather more than less. Therefore, whenever some modelling seems to work, a whole throng of rigid auxiliary conditions and similarly rigid orthoregulations is to be expected tinkering about in the background, be it in form of decora, hidden codes, laws, or social conventions. Consequently, catastrophes in the global financial system are a clear sign of freedom, and similarly the jealous endeavours of the EU political class towards "stabilizing the markets", as they say euphemistically, are a no less fallible harbinger of maximized bondage to come.

And the third reason, finally, for the impracticability of the model accuracy question lies in the fact that any model is able to prove its workability only a posteriori, and therefore its accuracy only in the future. Accuracy cannot be established, if for nothing else then for the simple reason that establishing it would require an evaluation and consequently another model. Even once the future has arrived, i.e. a sufficient number of observations relating to a model have been made, we can still only say that the model up to that moment more or less proved that it is in sync with our Lebensform. The world itself is no lab table, nor is it placed on one. When we are modelling, we do so within the world, in order to connect to the world.

Thus models cannot be accurate as long as one believes to be pointing to an external reference, and that much less can they be true. Models are tools of anticipation under the expectancy of weak reproducibility, a constellation we often call consistency. We have no other choice, not even if, within a revelation-oriented religion, we were to try to forswear modelling altogether. But for rather limited meditations, even a mystic cannot do without models, and much less can the mullahs or the Pope, representatives of institutionalized revelation, except perhaps if experiencing a trance.

The space of *diagnostic anticipation* is alone in affording us a reference to the world. This also goes for all the cases and circumstances where we implement decora, traditions, or legal conditions in the guise of cultural programming, with a view to improving the probability of successful modelling. But models born into a strong structural frame are centred models. Either they are centres themselves or they refer to axiomatic dogmata. Such constructs are unhelpful in our quest for or examination of the possibility of machine-based episteme. In other words, there is no such thing as programmed theorizability or programmed smartness. Asking about the possibility of programmed smartness is quite as inadmissible as the question of whether this arm here belongs to me or whether there are mushy stones.

In short, and under a practical aspect, models must prove themselves in accordance with an individual, specific and most of all dynamically unstable constellation of purposes and intentions. Were we to introduce some optimality criterion, or to claim measurability of purposes, or mapping of purposes onto contexts, we would be landing ourselves straight in utilitarianism. We would then be speaking of cost functions and strategies. While that might be useful in individual cases, in a generalized form, however, I think it more than just an exaggeration. Because the criterion of optimality would, right again, mean reintroducing an ideational centre, a centrement, with all the counterintuitive consequences of other centralizations, i.e. axiomatizations. The striving for implemental optimality of a catalogue of purposes leads straight into non-securable relativism, or relativistic axiomatism. In other words, war.

If we forgo, however, the reference to optimization, we are, on the one hand, reverting to Wittgenstein's Lebensform (Forms of Life),[19] in a French variety, as it were. And on the other hand, we shake hands again with Leibniz and his wonderful concept of absolute inwardness, the monad. Because purpose or intention cannot indefinitely be tested for some advantageousness, nay optimality, or vindicate a strategy, as this would implicitly mean that we possess an all-verificatory model. Such a position quite obviously is unreasonable, even for formal reasons, but no less obviously are we likewise denied the possibility of corresponding empiricalness of infinite verifiability. The posing of a purpose hence recurs to some thing that appears even before the person of the subject. Just what Leibniz calls absolute inwardness.

Having thus ended up at absolute inwardness, am I now finding myself back at the middle-point notion, with the concept of de-centrement having thus failed? Not at all. Leibnizian inwardness must not be confused with a middle point, the correlate of some quantifiable spatial extension. For Leibniz, monads are pure qualities; they imply absolutely nothing quantifiable. Nor is absolute inwardness locked up in some strongroom, as it were. On the contrary, it is alone to touch the world directly. "World" here shows up as the machinery of the body, or rather—once more with Guattari—as the corporeality of the machine-like with all its perceptions and affects, and—as per Deleuze—the seething of its parts and the sense paradoxically emerging from it. As for ourselves, we have no access to our own absolute inwardness other than through models, models about a part of ourselves, mind you—which, referring to Leibniz, Deleuze calls folds.[20] Yet, it is likewise true that without a will—which ultimately can only be collocated in absolute inwardness, as we have seen—the subject cannot model. In other words, the subject, the will and the model actualize and concretize simultaneously; to my mind they are the three first, indispensable and non-reducible parts of any *status nascendi*, but in particular of spontaneity. Hence, neither the subject nor the

---

19 Savigny E.v. 1995, "Bedeutung, Sprachspiel, Lebensform", in: Apel K.-O. et al. (eds), *Wittgenstein Studies* 2/95. Online: http://sammelpunkt.philo.at:8080/456/1/11-2-95.txt (accessed: 10 June 2011).
20 Deleuze G., *Fold: Leibniz and the Baroque*. University of Minnesota Press, Minneapolis 1992.

will are psychological concepts.[21] Putting the model into the role of a constituent of the subject also leads to a deep link between any of the potential or actual subjects. This linkage of subjects then enforces any problem related to any kind of "justification" (knowledge, truth) to disappear.

From this, there run manifold implications that I cannot pursue here. As in political theory, where human dignity is precisely grounded in the notion of absolute inwardness, seeing that it secures the dignity-constituent difference between within and without. Regarding our topic, machine-based episteme, we are thus getting a clear pointer that our thinking must move in categories different from cybernetic control and numerability. Conversely, this provides us at the same time also with a further argument against the system-theoretical, i.e. cybernetic conception of the subject, à la Luhmann or Habermas. Thus, Luhmann's or von Foerster's conception of second-order cybernetics reveals itself inherently as nothing less than a massive attack on human dignity. As an insight, this is not new, but the reason here presented—we are coming from pure modelling, are we not?—would seem somewhat interesting. By the way, with the help of Leibniz and de-centrement, radical constructivism can be refuted as well, because Glasersfeld,[22] Maturana and Varela do not simply reject performatively the very primacy of de-centred modelling—they utterly ignore it; in their thinking the link of the innerworld constructs to the world remains extremely opaque. At that, radical constructivism maintains that everything (!) that we are able to think does spring from our brain. How there exist such things as conventions, and why these conventions happen to be congenital with language in our world, Varela is at a loss to explain.

At this point, a misunderstanding must be pre-empted. I just stated that inwardness in its nature as pure quality was meeting the corpo-reality of the machinic. This might be construed as somehow on the way towards some dualistic substantialist ontology. That however is the last thing I would have in mind, for it would mean once more absolute separation of body and mind and thus strong centralization. The very same relationship we ourselves as subjects entertain with respect to our body is, of course, operative at all corporeal organizational levels and with all single entities down to the molecules, in relation to their respective constituent parts. In some way, absolute inwardness is not absolute; rather, it represents the ideate sphere that is shared by the transcendent and corporeality. It is absolute only in the sense that it is itself an immaterial, immanent a priori without any further premises.

---

21 Note that we do not need objectness in order to allow for the possibility of subjects.
22 Glasersfeld E.v., *Radical Constructivism, A Way of Knowing and Learning*. Routledge, London 1995.

Today, the notion of absolute inwardness is rarely used, although it may be, as we have seen, a tool for disposing of the body-soul problem. For total disposal, we would have to discuss symbols, which cannot be done here. The inwardness notion can be found, if at all, in the context of theological dogmatics, or of proofs of the existence of God and their critiques. These are not my topics. But the notion does indicate a point where modelling comes to an end. Wittgenstein moved this "point" into the exterior, into the collectively practised Lebensform, Leibniz however into the within. The two are closely related. Both have very little to do with religion or indeed the idea of God. This contradiction is easily resolved, as I shall demonstrate in closing. But first, let us turn to the practice of modelling.

V THE PRACTICE OF THE DE-CENTRED MODEL

After having, up to now, theoretically de-absolutized the model concept, as it were, i.e. de-centred it, I shall now demonstrate briefly, and as concretely as possible, how de-centred modelling may be practically formulated without becoming recentred again by such formalization. The model must ultimately reveal itself at once as living in a structure and being one—one that is its own sufficient cause. The strong implication derived from this may perhaps sound scandalous, indeed, placing as it does formalization and creativity as co-conditional into the anteroom of the very thinking itself. We must here forgo looking into this further, but would point out that we consider the ideas commonly attached to the formal as being in urgent need of overhaul.

As we set off towards formalization, we now certainly expect that on this basis de-centred modelling may be programmed, in so far as it is programmable at all. De-centred modelling cannot, of course, be programmed as a totality—there is no private de-centrement, as there is no private language. Conversely, our own corporeality, our own sheer materiality including mechanisms that keep that body going, are providing an excellent argument that de-centred modelling, too, should be mechanismable.

As we are, in this context of de-centrement and construction, inspecting modelling, we are implicitly formulating the model in the furthest-possible generalized, nay, in a radically generalized fashion. The general definition of a model must cover the potentiality of modelling, without however forgoing the link to its conventionalistic embedding. In other words, the general formulation of a model must admit of consciousness of modelling while modelling. Or put more linearly, we must keep in mind all possibly thinkable premises of modelling—at least the ones I termed "handicraft-like" alternative a-prioris above.

So let us visualize an imaginary or even demiurgic-aprioric lab table, onto which to lay out our modelling ingredients. What do we need?

For one thing, there are observations. To start with, we have to admit to being incapable of observation without any tool, organ or theory, whence the threat of either centralization or infinite regress. However, we shall see later how to get rid of these contingencies again. Yet prior to observing, we need a purpose or an intention. Mostly, we do not explicitly formulate such a purpose, we simply use models, e.g. because we have always done it that way. We impose a certain usage upon the model, stamping its purpose a posteriori upon it, as it were. Purpose, intention or usage may be arranged along very extended time scales and therefore be nearly invisible to us. That does not detract from their being indispensable for modelling. Negating the need for purposes produces one of the severest threats modernism is capable of: it makes us forget the rack, the racking and the related categories of virtuality and responsibility. In emphasizing usage we are actualizing Aristotle's final cause as well as Wittgenstein's theory of meaning. As we already mentioned above, purposes need not be spelled out explicitly, nor need they be identifiable. Any purpose, however, becomes operationalized at one stage of the modelling process or another. This operationalization unavoidably introduces a gap, for this mapping from purpose to operational goal can never be a complete one. Of course, this gap can be operationalized itself into the notion of risk, or more appropriately, into a component of risk.

With usage as a backdrop, we are able to sift out "features" derived from potential observations. That means we are evolving them. The entirety of an observation thus consists of actively selected qualities or properties. Let us take stock exchange prices. Technical chart analysis provides such secondary "properties". Another closely related example would be the perception of tempi in a piece of music. Putnam's proto-semantic notional extensions also belong in this context. The primary characteristic of these "properties" is their originating from an active selection of the measuring instance. This process, this activity should be seen as "propertization" as it were, as generation, and—strictly hitched to it—appropriation, of the objectness of what is being measured, in a sort of immobilization. We are props managers in the performance of our authorship. It seems as if the life-worldly practice of modelling inevitably were bringing about crystallization, fixation, territorialization, and correspondingly centralization too. If that were the case, we would now have to call off our attempt at describing the de-centred model. But it is not, fortunately. The impression of an implicit linkage between modelling and centralization is deceptive, and the cause of the deception sits in our stance toward habitude. One might nearly say the impression of deception is the cause of deception.

The upshot of that "propertization", of those activated "properties", are construed quasi-objects within the immaterial, frequently called qualia. Conversely, all qualia are liable to mensurative breaking down into

components. There is nothing mysterious about our visual sense extracting, right at the prime levels of stimulus processing, the "properties" of "red" from the stream of potential stimuli, i.e. light. We are then playing the language game "red", which to us does not seem to be further reducible. But this a fallacy, a sensation that is only true in so far as we were to consider our perceiving body itself as a—and as *nothing else than* a—non-reducible entity. But that would lead straight into shamanism. Hence qualia are not in any way absolute in a transcendental sense. They represent simply an ontogenetic a priori. The biologic, design-conditioned convention is certainly one of the strongest of them, but weaker ones, such as social conventions produce qualia as well.
So far we are dealing with a 3-tuple, to be thus symbolized:

1   $MO = \{U, O, P^o\}$
    U = usage, which includes the goal (purpose) and risk measures;
    O = potential observations;
    P = construed or selected "properties" imposed to observations by means of "propertization"

The usage part, furthermore, has an interesting and often-overlooked internal structure.

For all modelling, we are bound to operationalizing our purpose. This we achieve very simply through selecting or building a criterion variable. Within this criterion variable, there are several variants that are of particular interest to us. A building, for example, ought to withstand an earthquake; i.e. a certain value of the variable "tremor" ought not to be conducive to a change. Or in a medical treatment, a certain dose of an agent ought to produce recovery. In general, we are not interested in any global analytical description that would map out all imaginable influences upon all possible results. Furthermore, natural systems are for the most part highly non-linear, which precludes such mapping in any event.

Which leads us to one of the most important components of transparent, de-centred modelling: the semantic asymmetry of the consequences of its application or, in brief, the differential cost of errors.

Using models in the end is always about classification; action A is carried out rather than action B. Such classification may be erroneous; indeed it will in most cases be faulty. Now, there are two types of errors. Either the model is to recognize constellation A, but fails; it does *not*, by error, lead to the preferred action. This is called being false-negative. Or else, the model suggests unfortunately the preferred action A while circumstances do not bear this out. This is reasonably called false-positive.

Wrong decisions generate costs measurable, in generalized form, in terms of time or energy. These costs are apt to be dramatically asymmetric for both types of error. And being dramatically so is the rule. If a doctor tells me erroneously that I am well when I am not the costs can be disastrously high for me. The opposite case is much cheaper, which is one of the reasons for our health cost problems.

As seldom as this asymmetry is at all recognized in technical modelling, as varied are the actual models that result from a particular choice regarding the error-cost relation.

So, we put the usage or purpose of a model into symbols as follows:

2   $U = \{\, TV,\, TG_{TV},\, ECR \,\}$
TV = criterion variable, basis of mensurative operationalization of purpose
TG_{TV} = target group, defined as subset of TV (e.g. as interval)
ECR = error cost ratio, i.e. the relation of false-classification costs of alpha or beta type

There are further, less momentous dimensions of usage, which will be ignored here.

We shall now complete the first 3-tuple (f.1)

$M_\theta = \{\, U,\, O,\, P^o \,\}$

from the more evident parts of a model with three additional, frequently neglected but no less essential components.

They are:
the algorithmic procedures employed, A^p
the particular conceptualization of similarity S^c
the symmetric properties of an implied quasi-logic Q^s

The first of these elements is directly understandable. Procedures describe, for example, how a difference gets ascertained, which might be programmed in some computer language or set down in a manual. Here are buried all those thousands of methods, from statistics to hermeneutics, together with their specific implementations. The element A^P reflects the fact that it does matter how something is being performed. A difference must be made between it, and the operationalization of similarity going to be applied.

Similarity of two observations is ascertainable, i.e. operationalizable and thus calculable, in various ways.[23] It is obvious that it is nonsensical in modelling always to apply precisely just one specific mode of similarity calculation or even aprioristically to imply its appropriateness. After all, it ought to be up to usefulness to decide on the choice of method, in the sense of successful anticipation, and not the other way round, up to the method blindly to restrict usefulness. In other words, the particular concept of similarity chosen for some specific modelling depends on the purpose and on available data.

---

23 Any mapping (in the mathematical sense) of some observed values onto a comparatively smaller set of some other value(s) could be regarded as a particular similarity function. Despite the fact that there are many different possible ones, in many (if not most) applications Euclidean distance is used as an operationalization of similarity. Yet, any statistical procedure, e.g. correlation, also actualizes a similarity mapping. c.f. Blough D.S., "The Perception of Similarity", in: Cook, Robert G. (ed.), *Avian Visual Cognition*. 2001. Online: www.pigeon.psy.tufts.edu/avc/dblough.

Besides, the method of similarity ascertainment ought to resort to the fewest possible further conditions. Statistics, for example, which means just any statistical method whatsoever, demand independence of all partial observations, i.e. independence of the properties of the measured system. It is this fundamental and otherworldly requirement which, as an axiomatic precondition, renders the whole statistics field so questionable. But as soon as this premise is abandoned, a numeric method ceases to be statistical.

Now for the last element of the generalized model, the symmetric properties of the applied quasi-logic. This includes, for example, the necessity that a particular modelling must in the first place find out whether a linear method is to be used, whether observations should be mapped into a self-referential mechanism, whether to use global-geometrical or rather local-topological approaches.

Then, prior to modelling, a logic must be chosen. Independently of similarity operationalization, or method, a specific logic is needed, e.g. for aggregating intermediary results, which may well be contradictory. The classical Aristotelian bivalent logic in practice often produces intolerable artefacts. This is why multivalent logics are used as a rule, or even logics on a t-norm, i.e. logics with an infinite number of grades between true and false.[24] Then, the choice of a certain logic, just like the other parameters, may be dependent upon a context or a course of proceedings.

The symmetric properties overlying a quasi-logic are thus in their turn a compound,

3    $Q^s = \{ Loc, Lay, Lin, SR \}$
Loc = locality, dependence on the process of distribution formation, context sensitivity
Lay = capability of self-constructive stratification into conceptual (epistemological) layers
Lin = scale-capable linearity (incl. commutativity, associativity)
SR = self-referentiality

In the aggregate, we thus obtain the following tuple for formulating the maximally generalized model.

4    $M = \{ U, O, P^o, S^c, Q^s, A^p \}$

Now we may turn to practical implementation of this general model formulation, meaning that, as presented here, it can act straightaway as a guidance for meta-programming. Actualizing this differential model into, say, a software for modelling, is more or less independent of the choices for particular methods, even though some may be more suitable than others.

---

24 Gottwald S., 2009, *Many-Valued Logic*. Online: http://plato.stanford.edu/entries/logic-manyvalued/.

Here, we would like to mention very briefly a critique often met when presenting such symbolizations to mathematicians or physicists. Usually, they claim that such symbolization should be regarded as meaningless, since it is not accompanied by any kind of "operation", or transformation. It is true that f.4 does not represent a calculus (yet). Nevertheless, it immediately triggers questions about its structure. The symbolization introduces a certain detachment from the reference to the content, hence the question appears of whether the elements are equally relevant with respect to their deletion. What are the consequences of deleting one or more of them? What happens if we combine a complete model with a deficient one, and is there any regularity on the side of the operators in doing so? Is there a relationship between f.4 (or one of its subsets) and the definition of a mathematical group? All these questions (and many more) would not even be visible without the symbolized form of f.4; hence we consider f.4 as a well-justified starting point for formalizing the concept of model in non-reductionist way.

This construct as represented in f.4 above contains a fair number of parameters, orthoregulative parameters, that is. This is fine, as it not only allows all fixed points to be dissolved, but also the relevant to be distinguished from the less relevant modelling determinants. This way, against the backdrop of the de-centred, differential model, the whole method-related jiggery-pokery vanishes, in which not only science disturbingly has indulged itself for well over the past 80 years.

As these fixed points then get actually and practically dissolved, the above-mentioned formal representation will not be describing a model, but clearly the differential of a model. We need an instantiation that, as every instantiation, lies without. As per Wittgenstein, we might say, somewhat undifferentiatingly, it derives from the embedding Lebenswelt.

Regarding modelling practice, this means that never just one single model is created. Quite the contrary. An experiment will be performed, and in its course multiple test series will be established, and this experiment is not in relation to the world, but examines the influence that the instantiational conditions of the differential model exercise upon the representation and screening of the available observations, as related to the purpose. Typically, based on one and the same mass of observations tens if not hundreds of thousands of different instances of the differential model are established. In the end, many models may result, divergent regarding their use of variables, namely, their pointing to the world, but with similar usefulness. Hereby, true comparison of equal values is achieved; monocausal explanations, on the other hand, will be getting thin on the ground.

Put differently, instead of postulating, in the manner of a shamanistic expert mode, one single model, as the positivistic prescriptions would impose on the world,²⁵ the differential model provides us with a comparatistics. Its theme is not the uninterested interpretation of things past, but on the contrary the anticipation and structuring, many-layered and tainted by subjective intentions, of the future.²⁶ Conversely, we have now, too, a formal reason for the superiority of comparatist approaches. The differential model may therefore be relevant for many discussions in the humanities, too. It removes the at once opalescent and opaque covering layers from the rituals staged around diagnosis and prognosis. From the differential model, as a radical de-centring of the relation to the world, it becomes clear that every one of our decisions is about an infinity of possibilities. This phenomenon we call liberty. Liberty, thus, in no way is a craft,²⁷ and cannot, and need not, be contrived. Every relation to the world is soaked in liberty, as it were. I would even say that the claim that liberty must be generated conceals a malicious ideology. Incidentally, this ideology calls itself "analytical". But the contrary is true, actually. We have no other choice than to configure and negotiate liberty for each and every pronouncement. And this is precisely what is missing from analytical perspective. Liberty cannot be an ethical value in itself, at the same time as it enforces ethics and is always subordinated to ethics.

In addition, the differential model dissolves a further ancient-scholastic duality, that of object and subject. Within our framework, it is absolutely not the case that there may be talk of subjects only as long as there is of objects too. The differential model replaces the object, or the coherentist theory of the object, including any attached substance ontologies and representationalist phenomenologies. For all that removal, the subject does not stand isolated; differential modelling is its counterpart, in other words an environment made of differential models. Differential modelling, of course, cannot do without conventionalistic notions, embedding Lebensforms, nor without recourse to mediality or to virtuality. The accommodation of the subject within the world thus changes radically, in contrast to the all too simplistic subject-object scheme.²⁸

VI THE NON-ASCERTAINABLE LOCUS

Now let us carry de-centrement to its completion, in support of our Weltbezug via a differential model. To that effect, we have to get

---

25 Statistics actually prescribes even fixing the test before the measurement, and then doing measurement as well as the statistical modelling only once on a given set of data. Without that restrictive methodology the maintenance of basic assumption of logical empiricism would be violated, such as the independence of variables and data points.
26 In some way, a trans-Kantian anti-Kantianism, at least regarding the possibility of a "foundation" of reason, i.e. a priori knowledge.
27 As the analytic philosopher Peter Bieri tries to argue. Bieri P., *Das Handwerk der Freiheit: Über die Entdeckung des eigenen Willens*. Hanser, Munich 2001.
28 As it is favoured for instance by Slavoj Žižek and other belated political materialists. Žižek S., *The Sublime Object of Ideology*. Verso, New York 1997.

rid of all apriorisms as well as all restraints by whatsoever methods. We shall have to succeed without forswearing models and their premises, i.e. worldliness. By the way, it is precisely the negation of the model that makes the great revealed religions what they are. Whenever we reject as subjects the need or the possibility of some modelling, we imply a revelation-based construct. The inverse holds true too: the idea of revelation itself is not particularly problematic, as long as one is not pretending to be able to express something about the world by means of it, because that would once again be mixing up language games (relying vs. denying modelling), and violating the self-imposed rules, things that without exception produce strong, if not catastrophic consequences.

Doing away with all premises seems to be impossible. So the unavoidable fact of conventionalistic embedding remains. We are in need of signs for modelling, even for implementation of meta-programming, and they are indeed subject to conventions. Conventions themselves often do mutate, via the detour of rigorous standardization, to a sort of programming language, but this does away with the unavoidability neither of conventions nor of notions.

If we expose ourselves to this duality of notion and model, and then pause for a minute, we recognize immediately that the notion as well as the model must be counted among the transcendentals. There is no pure notion, nor pure model. Two epistemic transcendentals having now been identified, the question arises of whether there are others, and how many of them would be required for sustaining the possibility of episteme at all. It must be remembered that within our de-centrement programme we are denied the dogmatic, i.e. axiomatic solution for breaking the impending loop.

Preconditions can be got rid of in two different ways. The direct one obviously does not work. For us, following the programme of de-centrement, preconditions cannot be jettisoned just like that, axiomatically. So we have to turn to the other one, we have to formulate the team made up of model and premises in such a fashion that both model and premises are being berthed within the model itself.

To this end, we should construe a specific space. This space must not have any ascertainable middle, nor must it have any ascertainable locus at all. As paradoxical as it sounds, there is nothing particular at all about such a space.

Think of the Escher discs with the dragons, the butterflies, or the fish. If we now imagine such a hyperbolic space[29] as being a differential space—i.e. there are no x, y coordinates here, only their derivation—then here is our space. Differential space exclusively contains motion. Any attempt to settle in one place or identifying one, would instantaneously find itself in another place, on the sheer strength of the identifying.

---

29   Anderson J. W., *Hyperbolic Geometry*. Springer, New York 1999. Chap. "The Poincaré Disc Model." §4.1, pp. 95-104.

Now notions and signs are not the only strong preconditions for models to be of similarly transcendental nature. In our view, it moreover takes virtuality and mediality. Only with these four aspect-like dimensions is it possible to construe a self-containing space for episteme.

The said four dimensions, however, are not in an orthogonal arrangement to one another, such as in Cartesian space. Much rather, every change in respect of one of the transcendental poles is accompanied by a change in respect of the other poles. At a first instant of figurative representation, at first sight as it were, one might believe that one sees a tetrahedron, where the transcendentals virtuality, mediality, notion and model each occupy one of the four corners. But neither the four corners of this figure nor the edges are "existing" in the sense of "ascertainable". Both must be thought of as being relative in a similar, yet at once more radical manner, as Descartes demonstrated with the "origin" and the axes of his coordinate system. Our quasi-tetrahedral figure therefore may not include any flat Euclidean space. The formal analogy, as some operationalization of sorts, of the requirement of transcendental poles is hyperbolic space. In our transcendentally delimited space, and regardless of "delimitations", there exist only directions, but no end-points.

Thus, any particular visual representation could be deeply misleading. The described space does not even have well-defined borders or "planes", not even in a topological sense. Using the tetrahedron just serves the purpose of indicating the four transcendental poles.

Once more, any muddling of language games must strictly be eschewed. As opposed to Cartesian space, ours does not contain co-ordinates, either. It is utterly impossible to "point out" a place (fest-zu-stellen) or to occupy it. This is easily comprehensible when observing what happens as one begins to think about something. As soon as thought becomes concrete, we must bring in models. But every use of models is simultaneously attended by surrender to notions, it continuously generates potentials, which are expressions of a surface on whose opposite side we collocatively "think" the virtual. The moment we are thinking of a locus in this space, we are instantaneously "elsewhere", whereby this "elsewhere" results, or generates itself, depending on style and context. Anticipating this motion, in the sense of expectation, is utterly impossible.

Such a space seems exotic and paradoxical, to put it mildly. Yet this impression only holds from an Euclidean perspective. Such, even here simple operationalization is possible, an instance of a mathematical figure so to speak, of the second-order differential. Our proto-epistemic space contains no loci, just changes of vectors. There are no figures made of loci and points to be seen, just accelerations, modifications of movements. These figures are non-configurable, as an object might be configured. These figures stand for nothing else than the dynamic quality of a Lebenswelt. Therefore, we call them *choreostemes*.

When trying for a second to visualize some of those figures, the particular quality of that space becomes directly clear. Claiming stability of one point in such space would thus be tantamount to claiming universal immutability, better known as idealism. A while ago, we were already in touch with another figure, namely negation of the model. It is interesting to note here that every choreostemic figure that is assured of having the transcendent model pole's backing, feels like a (small) revelation, no matter how strong the model pole's emphasis just then happens to be. Revelations therefore are just as experienceable to a mathematician as they are to a poet, a logician, or a believer. Negation of models, conversely, can never be imagined for more than a moment, not even in complete religious trance. We cannot go further here into discussing the various figures. Let us just point out those that form between the (transcendent) poles of model and notion.

Augustine famously said he knew exactly what time was, as long as he did not think about it; but the moment he was beginning to think about it, insight was gone.[30] This statement as a whole may of course pass as pre-critical at best; however, it demonstrates with precision the nature of choreostemic space. Ideas are not communicable in self-contained form. Attempting to communicate the content of ideas makes no sense, and as such content cannot exist. Other authors, such as Peirce and Wittgenstein, from totally different angles, never tried to insist on that fact. The choreostemic areas extending between the notion and model transcendentals afford the possibility of relating all of the following, models as self-exposing anticipations, the world of symbols and symbolizations, logic, mathematics, semiotics, each in its turn, as Lebenswelt practice and imagined notion in such a manner to one another, that the structural-genetic infection of logic by the world, together with discoursing about it, may succeed. Thus, we have demonstrated how choreostemic space, in its quality as a special amalgam of some very few transcendental entities, helps us to imagine the abstract genealogy of notions and ultimately of language. In this, choreostemic space functions also—even quite literally—as a translation machine between choreostemic figures and fields of various Lebensformen and cultures. In almost rule-like fashion, abstract ferries may be construed that facilitate visiting other constellations.

And now, what does it all mean for machine-based episteme?

First of all, every instance provided with epistemic capabilities must be able to move in this space, if it is to function as an epistemic instance at all. It must be able to model autonomously, to manage the models and

---

30   Augustine, Confessiones XI, 14. "For what is time? Who can readily and briefly explain this? If no one asks me, I know: if I wish to explain it to one that asks, I know not: [...]" cited in: Modell A.H., *Other times, other realities: toward a theory of psychoanalytic treatment.* Harvard University Press, Cambridge (Mass.) 1990, p. 76.

to employ them differentially. This includes expositive relationships as well as the need for ascription of a personality. In turn, this presupposes that there cannot be any higher instance, such as an external programmer, who might decide how this movement were to instantiate. Following a rule of course does not mean being a rule. The epistemic instance thus must be autonomous.

And finally, it must be capable of classification and of symbolizing it vis-à-vis its own exterior. This entails virtuality, mediality, and Schleiermacher's hermeneutic circle. Schleiermacher wraps it up his own way when he embraces, for text and the work upon it, the "individual and the general".[31]

Radical modelling, consistently seen against the foil of machine-based episteme, thus turns out to be perhaps not just a link between the subject and the world, but a bridge between cultures that, under the influence of reductionist views, have drifted apart over the last 150 years.

INDEXICAL MARKINGS OF THE TOPICS DISCUSSED

In a rather indexical and compact form, we would like to provide some coordinates of the issues discussed after Klaus Wassermann's lecture "That Centre-Point Thing. The Theory Model in Model Theory" at the Printed Physics Conference, and that is presented in this chapter as a manuscript with some slight revisions.

A first line of discussion focused on the difficulties involved in finding a sufficiently abstract position for a theory about modelling that would not have to limit itself by specific metaphysical, ontological or epistemological assumptions. In the way these assumptions are reflected in different treatments of the transcendence / immanence scheme, they are distributed through the history of philosophy. The claim from Wassermann's theory of being able to break out of this scheme altogether was discussed. It was argued that this cannot be achieved by seeking the proximity to monad thinking (which Wassermann had

---

31 Frank M., *Das individuelle Allgemeine: Textstrukturierung und -interpretation nach Schleiermacher.* Suhrkamp, Frankfurt 1985.

emphasized). Leibniz was unable to break out of this scheme with his attempt at thinking of total immanence. In fact, he had to refer to the acceptance of a pre-stabilized harmony between physis and monad and therefore remained caught in an absolute dualism. However, Wassermann said that this dualism with Leibniz was linked with the solipsism of his monadological thinking and that his own concept of a radical a-centralization should open another path. His presented model of this space (in which one models) claims neither ontological existence nor metaphysical ideality, but rather should itself be assessed as a model. The central idea of his modelling theory exists in providing a comparatistic approach to the search for how models can be anchored. The central idea is to deal with the problem of anchoring models not by searching for a plane of reference, but rather by constructing one in a comparatistic way. During the discussions, other reference points were sought after in philosophical discourse that would be suitable for reflecting and discussing the presented model. For instance, the Deleuzian reading of monads was cited, which tries to understand the monads as the being of becoming and introduces the concept of planes of immanence in order to avoid the impasse of an absolute immanence. The extent to which Deleuze himself remains caught in the old scheme of transcendence / immanence was discussed. The discussions concerned whether Deleuze was actually successful in achieving another space for the logical treatment of statements with his strategy of radically putting the method of infinitesimal analysis into an algebraic context. An additional reference that was considered was that of Hans Sedlmayr and his views in *Der Verlust der Mitte* (On the Loss of the Middle). Based on this, the discussions about the presented model space turned to its relations to the modular systematic thinking of modernity. The reasoning that Wassermann's model space allows training is not so much directed at a modular assembly, so the shared conclusion in the discussion, but rather allows for an increased capacity in integrating highly differentiated elements into one's modelling. A theory about modelling as a theory does not refer primarily to doing, but rather to the thinking behind the doing, and would be misunderstood if reduced to any specific catalogue or recipes for modular construction.

Based on this, a second central line of discussion was unwound by the relationship between modelling thought structures and the utilization of such thought structures for concrete applications, such as in architecture. In the discourse of architectural theory of the second half of the 20th century, as was in one argument brought in against the abstractness of the presented theory, there are numerous contributions by Tschumi, Libeskind, Eisenman, Lynn or Novak, among others, that postulate such utilizations of theoretical thought structures for concrete applications and suggest mappings for them, concrete models that, when considered conceptually, are supposed to express in one way or another such abstract notions as mediality, virtuality, hyperspace or similar. The question raised was whether one could not better illustrate what this model suggests here in other, much better "digested" thought structures—whether, in short, there is a need at all for one more, new model. A discussion arose around this objection that this model here does not want to describe a representational mapping for specific applications, but rather ultimately wishes to describe the abstract genetics or genealogy of the theoretical constructs behind such concrete models. It is about learning to understand better the complexity of modelling itself, beyond their utility for mimetic application.

An additional line of conversation eased out of the empirical relationship between the model, experiment and modelled phenomenon. It was discussed that models per se actually never can achieve what we impose on them. A certain doubled completeness phantasm was discussed as being particularly problematic, which carries an odd symmetry between positivism and religiousness. There was an immanent tipping point toward totalitarianism, which may perhaps also appear in this model. It concerns setting the initial parameters when modelling, and the valuation involved thereby. This argument was countered with the fact that evidence and disclosure ought not to be negated here, but rather can be reflected through modelling, as per specific dynamic form of thought while investigating. Therefore, to a certain extent, a critical attitude can be taken precisely because it is not attempted to search for each sphere separately. At that point, the objection was raised that this is very abstract and a practice could not be imposed. This objection was countered with the fact that, firstly, the structure of the space in which this modelling takes place is not determined, but rather is considered as a differential space always to be instantiated in specific ways for each situation first. On the other hand, the strength of this model could be seen in its ability to inform its own implementation as software, with the support of technology. Thereby it must not be forced, but rather may be offered. As such it would simply have to compete with other offers, in terms of enabling well functional, interesting and primarily unexpected perspectives, solutions or products for any complex domain of expertise.

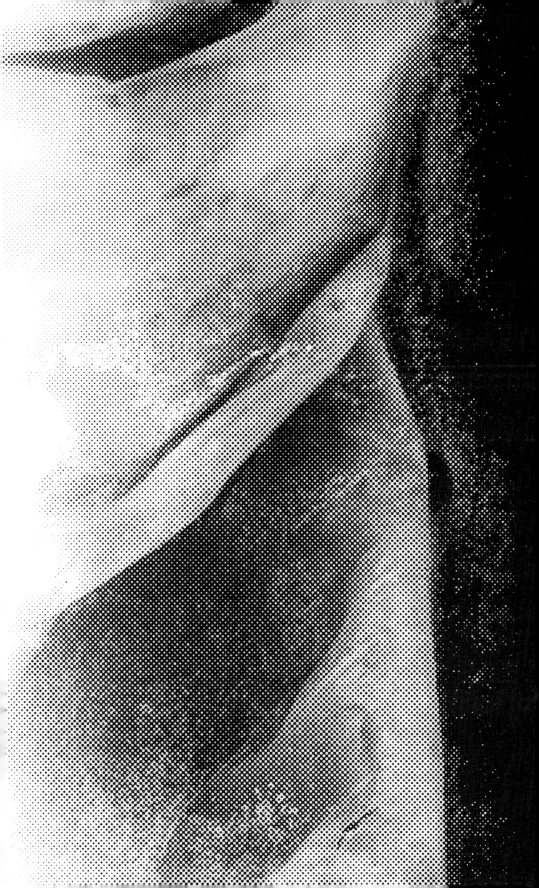

V DIGITAL CATHEDRALS—A FEW REMARKS ON THE QUESTION OF APPROPRIATENESS WITHIN ARCHITECTURE
HELMUT GEISERT

HELMUT GEISERT studied architecture and building history at the Universities of Technology in Karlsruhe and Berlin. He has conceived and organized several exhibitions on architecture, art and culture, and is author and editor of a range of publications in these fields. At the Landesmuseum für Moderne Kunst, Photographie und Architektur in Berlin he is responsible for the architecture collection. Geisert is professor at the Kunsthochschule Berlin Weissensee.

This article offers as food for though regarding the book's thematic phenomena of "printed physics", a retroprojection into the 19th-century continental discourse on the relation between the art of printing and architecture, or more precisely, on how printing techniques have then been received to impact on the societal role of architecture. Taking the risk of an almost inevitable romanticism involved in such a retroprojection, the article nevertheless makes a case for not forgetting about this discursive cultural heritage when trying to reflect on possible impacts of digitalization for a novel tectonics. We are, after all, only yet beginning consciously to experience and comprehend these impacts.

I would like to offer some thoughts about architecture and an architecture-related construct that had come to my mind and that might come in handy to the discourse of this conference as a parable. The following will not take the form of a closed argument in any systematic sense, but merely present some food for thought. Because, among other things, I do not believe that there can today be such a thing as development of a conclusive architecture-related system. That would be presumptuous, plus counterproductive after all. Surely, there is that claim that we are living at a turning point of times and that thanks to the digital technologies all things will henceforth be different. That may be true, but then we should be given an explanation of how that era, provided it came, might be apprehended.

Now there is the fact that there existed a turning point of times once already, the one leading from architecture to the art of printing. That view, if possibly a retroprojection, has nevertheless a lot going for it and may productively do service as a concept. It means that there were actually epochs when architecture did hold a front seat in human culture. So there would be one period, up to the building of the big mediæval cathedrals, and one with no cathedral building any more, cathedral building thus being the determinant.

One might now ask, pointedly, is the computer a cathedral? It is recognized, and illustrated in a beautiful text by Victor Hugo,[1] that the art of cathedral building, which embodied the entirety of human capabilities, was supplanted by the art of printing. So the era thus ushered in, which today in its turn is perhaps on the way out, would be the era of the printing art. And historically seen, this printing era then did manage to wipe out the art of cathedral building, thereby decisively recasting architecture's position. And further sharpening the question of our parable: is the computer a cathedral, or is it a printing press?

The idea of throwing a fantastic projection upon architectural objects is already extant in Victor Hugo's book about Notre Dame, as we have seen. The cathedral is taken as quintessential of human capabilities. Architecture is being presented as history of humankind's open book. "The cathedrals show variety, progress, originality, richness and resistant variety. From this folk art," so Hugo, "come only edifices for which no initiation is required to gain understanding of them. Every intelligence can understand them, every heart grasp them, every imagination plumb their depths, for its symbols are as easy to grasp as Nature itself. That is the Cathedral," he summarizes. "All that changes with the invention of the

---

[1] Victor Hugo, *The Hunchback of Notre Dame*. Translated by Frederic Shoberl, Philadelphia 1884. All paraphrased quotes in the following pages refer to statements by Hugo in the fifth book, yet they rely on different editions and translations.
Where not otherwise stated, the quotations in this article have been translated into English by R.R. Fischer. Due to the extensive and narrative usage of quotations, and to help support the character and the flow of an orally presented manuscript, which is the intention of the author, this article refrains from adding the original passages.

printing press. With it, the representation of knowledge and ability, and also fantasy, go over into the printed book." Thus Hugo's quite plausible (if loosely paraphrased) construction. "In order to understand fully the drastic impact of printing on the human mind, he continues, one must never forget that, up to the 15th century, architecture was humanity's main book. That up to that point in time, not one significant thought was thought that did not find its expression in an historic structure. That nothing happened that was not written in granite." Seen from this angle, that cathedral-vs.-printing plot is potentially a negative dialectic of architecture of sorts, as it socially hypostatizes the positive development, while being tantamount to its destruction. Unless today, one prefers to view this dialectic as a door to retrieving the cathedral's potential from the debris of the probably already consummated demise of the printing art.

That might well be our chance to set down a new epoch-making turning-point in keeping with our subject. However, such a thought is probably still premature today. But it would in any case be an extension of the architecture-defining construct as we know it, for example, from Bloch, where it is based on the contrast between pyramid and cathedral. Adding the printing art to those two, with a negative sign, one might be onto something. But reverting to Victor Hugo: "The printed book was called upon to obliterate the architectural monument."

The printing art of course introduced entirely new qualities into the divulging and establishing of ideas. Might architecture hold the same media-like qualities, such divulging and establishing of ideas? That would indeed be a rather new idea for architecture, since architecture was mostly perceived exclusively as a thing or an image. "Printing confers a heretofore unattained irreconcilability upon the thinking and lends it wings unassailable, irrepressible. It combines firmness with speed; from durability it achieves immortality," so Hugo. That means that the societal effects of printing are much more perdurable than the ones of any cathedral or architectural monument might ever be—thanks to the element of circulation, or rather its facilitation.

"Although the thought is only printed on paper, it is yet in its way more indestructible than stone," according to Victor Hugo (loose paraphrase). And furthermore: "The thought expressed in architecture needs a number of supporting arts." The next relevant point will be the division of labour, which was alone in having socially enabled that development. As Victor Hugo says, "Tons of gold, whole mountains of stone, forests of beams, hordes or workers were necessary to accomplish this result. The printed thought on the other hand needs but a little paper and printer's ink, and nothing further." The next question therefore is that of the consequences for architecture. The overcoming or, if you like, the killing of the cathedral through the advent of the printing art was likewise the starting point of the emancipation of the arts. Henceforth, there was going to be specialization, division of labour, fragmentation of the whole of work into single disciplines.

That implies at the same time that rules dissociated themselves from tangible architecture and became abstract. From this in turn grew all the constructs of architectural theory, and the foundations of plan design. That which heretofore was encapsulated in the art of cathedral-building becomes specialized and abstracted, so as to become communicable through print, thereby begetting treatises and—not to be ignored—adequate to the advent of civil society, which is indeed based on that destruction and that development.

Victor Hugo dryly remarks, "Without Gutenberg, no Luther." Hereby, however, the idea of cathedral architecture was sunk, or at least architecture's societal capabilities were fundamentally questioned. The outcome, according to Victor Hugo, is in the first place that "Architecture, like a pathetic beggar, trudges from imitation to imitation, and the greatest genius of the 16th century, Michelangelo, dedicates to her a last thought of desperation by piling, like a titan, Pantheon upon Pantheon in order to create St Peter's Basilica; this monumental edifice deserves to be the last of its kind. For it is the last representation of a great art."

This, of course, is a romantic 19th-century retroprojection, but still quite useful as a construct, as it does catch some of architecture's further development, or at least some of its problems. It might be said that subsequent to cathedral building, architecture coalesced into classicism. Architectural form increasingly disappears from buildings. In its stead, and concurrently, the geometrical construction plan gains ever in prominence. That means that henceforward in architecture we are dealing with real abstractions which, as we all know, are apt to be deadly. Hegel says in his Science of Logic, asserting abstractions as reality means destroying reality. And Hugo, "As in an emaciated patient the skeleton, the skin and bones, that's the whole art."

That reminds me of Aldo Rossi who, in his fundamental attempt to recapture architecture, tried to reduce it to its skeleton, not implausibly, seeing that the number of basic elements in architecture is not greater than that of the bones in a living being's skeleton. That's how it is, whether we like it or not, nor is there any denying the fact that in our modern age, architecture was indeed abstracted down to its skeleton. This is not a critical judgement, but simply an essential starting point.

Hugo finds that "Architectural art, however, is dead. Deceased without hope of resurrection. Smashed by the book, for being too expensive and not lasting enough." That holds true not only for cathedrals. If we contemplate what was built in the 20th century, and what is still being built, it may well be justified. Bruno Taut, in his treatise on architecture[2] published in 1937 in Turkey, in my eyes one of the best texts on architecture, states tersely, "In the first half of the 20th century, there was no architecture."

2   Bruno Taut, *Architekturlehre. Grundlagen, Theorie und Kritik aus der Sicht eines sozialistischen Architekten.* Edited by Tilmann Heinisch and Goerd Peschken, VSA, Hamburg 1977.

Victor Hugo, however, provides a beam of hope I should not wish to leave you without. "Perhaps in the 20th century (from the perspective of the 19th) there will occur the stroke of luck that an architectural genius will be born, just as in the 13th a poetic genius, Dante, appeared quite unexpectedly."
That of course is an outgrowth of the 19th century's cult of the genius. But we also know that, when qualitatively reviewing 20th-century architecture, there always crop up three to five architects, the perennial question then being whether with Gropius or without Gropius. But seen from Hugo's perspective, the thought is certainly prophetic.
There is then another meaningful remark of Hugo's, pointing to the evolution and the qualities of that abstract momentum within the art of printing, to the effect that "the encyclopædia is humanity's new Tower of Babel." Which brings us back to that point where, slap, the very results of printing demand to be apprehended through architectural notions. And surely we may say that in our manner of perceiving architecture we are still looking at something of a "Tower of Babel". Thus, the architectural art and the printing art sum up the cultural capacity of humankind. Architecture's capacity possibly fades over into printing art, with dramatic consequences for architecture. It breaks up into single disciplines. And that's how the architect's role must now be understood. That's probably what distinctively marks today's architect: functioning under the duress of keeping the decay going, in order to be able to turn it around. It is no different in other arts. Take Schoenberg's music. There too, you will find that reversal of those labour-dividing upshots, that breaking-down into single disciplines is to be achieved within one person and its work. Which of course may be done only fragmentarily.
In architecture this is maybe no different—the results are easily visible. The architect must deal with the fact that now construct and form—that which Werner Oechslin termed the style-cover and the core—are irrevocably dissociating, sundering house and city as well in the process. This means that today there is no more naively saying that the house constitutes the city and that the city is made of houses. Things have become much more abstract. Nowadays, everything is about specialization-based processes. They are the prime upshots of the progressing division of labour and its attendant specialization. I shall be coming right back to the division of labour, as this is a very important point for architecture.
Put pointedly, this means that the architect, a supposed genius, is called upon to roll back if not undo those results of the division of labour. But these processes are objective in nature; ultimately, they materialize in adequate reproductive processes under the dictate of economics—economics of labour, of space, and of functionality. Nothing remains naïve. Basically everything is sucked into the whirl of economy, not in an economistic sense, but in one of space economy. One no longer thinks of space nowadays, one measures square metres. Consequently,

architects have for the last 50 to 80 years abstained from designing any interior rooms. In the circumstances, those did not need designing any more, all it took was multiplying two figures, there appeared a number of square metres, and that was it, as far as the client was concerned. Theories of form were out. Incidentally, they became useless in an objective sense as well, since there is nothing for them to contribute to up-to-date design processes.

The 19th century however still had form theories, in which a window is defined, and how to proportion rooms. In the 20th century, there are no longer windows, just surfaces and openings. Which is fundamentally different. It might look like a liberation, but in the first place it appears as a loss.

So it may be said that aspects of architecture should be understood solely as aspects of communication. The consequences for the idea of tectonics are serious. Accordingly, tectonics was to have been one primary reflex against the societally assertive division of labour. And not a naturality, as which it is commonly painted, "something about loads and props". On closer inspection, one will find that the tectonist, as a synonym for architect, is the one to whom it falls to roll back that division of labour or make it reversible at least enough that a decent, coherent result may be achieved. But ultimately, there is no giving it a positive twist. I am now going to turn to the theoretician who probably went the most thoroughly and possibly the most penetratingly into tectonics, Karl Boetticher and his book *Die Tektonik der Hellenen*. Astonishing insights can be found there, which have only little to do with the recovery of the Greek temple. In certain parts it develops into a theory of architecture that is extraordinarily captivating and will remain unknown to all those theorists who are content with reading Streiter. "Their having originated in this fashion explains the art forms' popularity. They contained an imagery that was understandable not just to art fellows, but, rather as a mother tongue, to the whole people from whose depth the 'tectons' had sprung."[3]

Here we meet again that urge that architecture be understandable in general-language fashion. It comes over as Utopia, and the supposed progress as regression.

"That notion's withering and finally disappearing from the collective mind is a fate that the language of art forms shares with the living language."[4] Thus Boetticher discerns architecture's loss of linguality, rather as Hölderlin formulates it in the Hyperion, "The language we speak, we do not understand any longer."

As for architecture, Boetticher notes that "The fact remains that in the absence of such analogical models none of the properties of one

[3] Karl Boetticher, *Die Tektonik der Hellenen*. vol. 1, Berlin 1874, p. 37.
[4] Ibid., p. 37.

component part's work form might have been brought to figurative representation."⁵ This means that for possible evolution of the domains seemingly left to architecture, we should look to analogical models.

So, decisive things are happening. Architecture is now obtainable only through models, and architecture suddenly turns image—the two are linked, a sign of the impotency of an iconic language.

At the same time, this may point to some awkwardnesses of digital forms of usage. The quandary stems from the fact that architecture in its digital aspect, too, has suddenly turned into nothing more than a description of surfaces, without getting structural. This then is the tribute it was thought was owed to division of labour, quite as in one of Loos' words, to the effect that he was wary of nothing in life so much as trying to invent new forms.

Boetticher, once more: "All purely figurative comparison, however, as the art form allegorically ought to bring about, was possible exclusively thanks to pre-existent models that allowed such comparison."⁶ Thus, architecture is just a reproduction of extant models. "Therefore, all art forms could not but move in circles of reminiscences of perceived objects."⁷

Just briefly regarding 19th-century architecture, it may then be seen as the era of categorizing, collecting, sorting, and at times—not always—trying to understand the capacity, the architectural capacity, of the history of humankind. That was less than uncomplicated at times. After a start with the Classical period, it already took several years for re-understanding the Gothic one, which was more complicated not only because of the structural forms not being readily comprehensible. Well, the 19th century had the ambition of creating some encyclopedia of the history of humanity, as it were.

And it may be said that the 20th century has not done much else than run that down again. Yet, nothing much seems to have been added to this field. Were Boetticher's arguments just taken as a justificatory strategy on behalf of classicism, some objective fact behind them would go ignored, which is closely tied to the constitution of civic society, namely, the loss of immediate perception—the next point to which I should like to turn. The loss of immediate perception as, once more, a consequence of the societal division of labour, such as it directly manifests itself in architecture. This means that reflection is wanted upon the relation between image and language, along with the recognition that mere reception of images evidently is legitimate only ideologically, and that such mediation towards the object is from the start impaired by perceptional modes. Architecture still cherishes the belief in the immediacy of perception and the collateral presumption of innocence.

5   Ibid., p. 37.
6   Ibid., p. 37.
7   Ibid., p. 37.

This also becomes apparent while working with CAD or with a derivation of forms from processes, a widespread fashion of recent years. "An initial perception independent of any background is inconceivable. Every perception presupposes, on the perceiving subject's part, a certain past, and the abstract function of perception, as a coming together of objects, implies some more occult act by which we elaborate our environment."[8]

So this mediation, which is detectable even in simplest forms of perception, along with the loss of any immediacy, is of course also a reflex to the division of labour and linked with a severing from the object.

Perhaps this difference between perception and object, as stated by Merleau-Ponty, was hardly experienceable at the building of the cathedrals, because craft and art were much less far apart then than now. And only along the decay of this unity, which co-authored civic society, abstraction takes form that lets architecture appear in a new light. Hence, the so-called ulterior development of architecture is prominently based upon the advent of labour-dividing processes and the attendant severance of architectural art from craft, with further effects upon the forms of perception. Merleau-Ponty says: "Spatial perception is not a particular type of state of consciousness or action, its modalities are the expression of the whole life of the subject and of the energy with which this directs itself, through his body and his world, towards one particular future."[9] There is also the corporeal aspect of architecture, particularly highlighted here, since the body is inexistent in classicistic theory, unless perhaps as a phantom limb.

One might thus speak of particularization of the perception of architecture, representing at the same time a multiplication of viewpoints. This is probably one of the main traits of architecture at the turn of the 19th century. One of its early expressions is the landscaped garden, which takes a centrally important place within the evolvement of the modern age, because it renounces central-perspectival space, which is tantamount to a multiplication of viewpoints. Superficially, it is attended by loss of architecture, but also by the evolution of new tectonics. Architecture based on classical tectonics probably cannot do without central perspective. During the late Baroque period, this centrement of perspective had already become obsolete, getting radically scrapped in landscaped gardening, where suddenly there is no longer any motional axis, no more than in architecture, it being replaced by multiple viewpoints. Schinkel, unexpectedly but not surprisingly, states in his treatise on architecture that, when taking his buildings in the park of

---

8   Maurice Merleau-Ponty, e.g. *Phänomenologie der Wahrnehmung.* Berlin 1966, p. 37. Engl.: *Phenomenology of Perception,* trans. Colin Smith, Humanities Press, New York 1962.
9   Merleau-Ponty 1966, p. 37.

Sanssouci as landscape architecture, one might remember that the quality of buildings primarily consists in their allowing of circumambulation. Which does away with this aspect of perspectival authority too, and with much of image authority. Also, one will remember that the development of modern architecture, be it in relation with Le Corbusier or with Loos, is rooted in the very fact that there is no longer just one single reading of architecture, but multiple readings, and that, not unlike cubist paintings, there are multitudinous ways of seeing rooms, and the harder one looks, the more rooms look back at one.
This, then, is what was quite plausibly explained under the "transparency" aspect introduced by Colin Rowe. A useful notion, but which does not explain it all. But this transparency itself in turn was bent, and relativized in the suction of economics, revealing that not all that space was needed when transparent rooms were an option: economization through fewer square metres.
These circumstances then became a matter for architecture, namely, the bringing of such economization into decent cohabitation with spatial notions. Would it not have been interesting for the history of 20th-century architecture actually to appropriate these elements of architecture and novel forms of perception, in order to develop something qualitatively genuinely new from them, e.g. the spatialization of layers, i.e. going beyond mere descriptions of surfaces, and actually generating spaces in the sense of spatial depth, to be created through piling spaces upon one another?
Superficially, this carries with it a loss of architecture or, more precisely, the evolvement of new tectonics, based upon the dazzling light of the Crystal Palace and no more upon stone-like monumentality. Under this aspect, the Crystal Palace, that structure of iron, and glass, and light, is of course the very take-off point of the new tectonics. This can hardly be overemphasized. Richard Lucæ, a Berlin architect, already remarked in an 1862 lecture that the interior space of the Crystal Palace might indeed be perceived as cast air. This, then, means that the architectural notion flips, and wholly new spatial ideas or spatial constructs may evolve in its wake. The idea of architecture mutates into one of autonomy, as prefigured in the so-called revolution architecture.
This, however, entails the loss of architecture's social consensus.
Boetticher was still a believer that architecture fundamentally poses and plausibilizes its own rules, that this is comprehensible in an educational sense—and now there is this conquered space of the Crystal Palace, with untold architects mesmerized in awe and unbelieving that thereafter the designing of architecture was still an option. That experience must have been overwhelming, not least because it meant the genesis of the idea, or even the reality, of a new tectonics, tectonics taking advantage of light, of colour (the Crystal Palace was not painted grey, but rather colourful, the iron parts red and blue), letting architecture dissolve into light, as it were.

This of course is the very point of architecture's being fundamentally called into question, that question which to me is unanswered to this day: is architecture still possible at all? Subject to being accused of nihilism, it might be sharpened. A society based upon the universality of exchange of goods tends to abstractions, i.e. loss of concretions. The shimmering light of the Crystal Palace thus would also be a reflection of the circulation of goods, a harbinger of a world that has for a long time replaced concrete relationships with the digital.

Nietzsche, known for a very radical grasp of the problems of his time, said in precise reflection of those new conditions of architecture: "Hereby, another type of human is indeed being increasingly disadvantaged and ultimately made impossible, first of all the great master builders. Now the constructive force is being stifled, the boldness of long-term planning is being discouraged, organizational geniuses will go missing. Who would still dare to undertake things whose completion requires thousands of years? The fundamental faith is disappearing, on whose strength someone is able to calculate, promise, anticipate the future in his planning, or make the sacrifice of believing that man has value only inasmuch as he be a stone in a big building."[10] Such is the societal situation that corresponds to the relationship between cathedral and printing art. But Nietzsche takes it one step further, since there arises a societal question, which he puts thus: "We are none of us societal material any more, that is the truth of the day!"[11]

That is the societal meaning behind this architectural evolution.

Let me now introduce a few texts pointing to new and important questions that arise for architecture from these points. First, the fact that the architect suddenly finds himself in a new position. He can henceforth no longer build on anything stable, as Nietzsche puts it, he cannot ground himself in society any more, because suddenly there are no longer any secure proportions. Proportion is one of the prime notions of 20th-century architectural theories that I deem acceptable and useful—proportion not only in a geometric or spatial sense. Indeed, in Semper's theory, expounded in *Der Stil*, it takes a central place; symmetrical authority is joined by proportional authority and, of course, "authority of substance [or content]", a category most important to Semper, which fact nowadays is merrily being suppressed when attempting to hijack him for traditionalistic purposes.

The idea of proportion reflects the fact that architecture, as a complex process, has a weighing role, as in determining the relation between usage and economy of spaces.

10   Friedrich Nietzsche, *Die Fröhliche Wissenschaft. ("la gaya scienza")*. 1882. Aphorismus 356. Kritische Studienausgabe, ed. by Giorgio Colli and Mazzino Montinari, Munich and New York 1980.
11   Ibid., Aphorismus 356.

This ultimately leads to the question of which spaces would be appropriate to human beings—once more the question about proportion. And to further questions such as: what is a construction that is adequate to the spaces?

Thus the architect is the one to whom it is incumbent—a probably impossible task—to solve this societal complexity, in edifice construction, building, or planning.

This thought was expressed very early on in Kant's *Kritik der Urteilskraft (Critique of Judgement)*. Quoting from the academy edition of 1908: "The *second kind* is the art of presenting concepts of things that are possible *only through Art*, and whose form has for its determining ground not nature but an arbitrary purpose, with the view of presenting them with aesthetic purposiveness. In the latter the chief point is a certain *use* of the artistic object, by which condition the aesthetic Ideas are limited".[12] And now the crucial part, "the suitability of a product for a certain use is the essential thing in an *architectural work*. On the other hand, a mere *piece of sculpture*, which is simply made for show and which is to please in itself, is as a corporeal presentation a mere imitation of nature, though with a reference to aesthetic Ideas; in it *sensible truth* is not to be carried so far that the product ceases to look like art and looks like a product of the elective will".[13]

This is a very precise, even the most precise definition of architecture I know of, and which is quite judiciously, if not projectable, so applicable in examining the relationship between cathedral and printing art.

This idea of a proportionality between means and ends too appears after all in most of the important texts on architecture. Such as in Jacob Burckhardt, in his *Ästhetik der bildenden Kunst (The Æsthetics of the Plastic Arts)*: "What is only important: that the truth of all forms act as perfect beauty upon the evident proportionality of the means applied to the task at hand; that necessity forms out as freedom; that general consensus could have formed about it."[14] Thus, all these subjects seem to have been spelled out with great precision as early as the 19th century. Such as in Friedrich Wilhelm Schelling's *Philosophie der Kunst (The Philosophy of Art)* from 1802–03: "Organic shape has a direct relationship to reason, because it is its closest manifestation and itself indeed just reason seen in reality form. Reason has an exclusively mediated relation to things inorganic, namely, through the organism, which is its direct body. Hence architecture's first relation to reason is just an indirect one and, being conveyed only through the notion of organism, a merely notionally conveyed relation. However, were [architecture] to be absolute

---

12 Immanuel Kant, *Kritik der Urteilskraft*. In: Kants Werke, vol. V, Berlin 1908 (Akademieausgabe), p. 322. Here the translations follow J.H. Bernard's translation from 1914 (the italics are Bernard's).
13 Ibid., p. 322.
14 Jacob Burckhardt, *Ästhetik der Bildenden Kunst*. Darmstadt 1992, p. 40.

art, it must live on its own and without intermediation through reason. This may not be achieved by the substance expressing any notion of purpose at all. For even in the most perfect expression, a notion of purpose, not having issued from the object, will not pass into the object. It is not the direct idea of the object, but of something different, extraneous to the object."[15] Schelling's idea is indeed that architecture is forever a repetition of itself.

Something similar may be found in Schopenhauer's *Die Welt als Wille und Vorstellung* (*The World as Will and Representation*):
"Therefore, the beauty of a building lies in the obvious purposiveness of each of its parts, unrelated to whatever exterior, arbitrary purpose of man (as such the building belongs to useful architecture), but related straight to the existence of the whole, to which the position, size and form of each part must maintain a necessary proportion, whereby if any part were removed, the whole would collapse. Likewise, the shape of each part must be determined by its purpose and its relation to the whole, and not arbitrarily."[16]

The idea of proportionality between means and end is also constitutive of the form of architecture. Karl Philipp Moritz treats this point in texts about architecture that to this date have found only scant attention. One of them, from 1789, carries the title "Is architectural ornamentation in the various columnar orders arbitrary or essential?" This indicates that all the problems of modern times were already quite present at the close of the 18th century, that they were intimately tied to the constitution of civic society, and probably suppressed later. All this was never recognized, although—or perhaps because—it already contains much of what Loos refers to in "Ornament and Crime", over a century later. So much for bourgeois historical awareness.

I am now quoting Moritz not merely because this is simply a very fine text: "Easiness of overall view is central to thinking, as it is in art. Whatever interferes with that easiness of overall view, be it per se and taken individually, as pleasant as it may, in art is as reprehensible as is, in thinking, idle play of imagination without relevance to the topic at hand. [...] In the same fashion as someone with clear insight into a matter treats it with greatest ease, where another fails despite his greatest endeavour, similarly only thanks to his easy overview of his subject is the artist capable of acting through art upon the human mind and captivating its attention."[17]

And finally, I am turning to Adolf Loos, because what he here writes seems to me crucial to architecture. At bottom, it is all about the question

---

15 Friedrich Wilhelm Schelling, *Philosophie der Kunst*. Esslingen 1859.
16 Arthur Schopenhauer, *Die Welt als Wille und Vorstellung*. 3rd edition, Leipzig 1859, p. 252.
17 Karl Philipp Moritz, "Versuch einer Vereinigung aller schönen Künste und Wissenschaften unter dem Begriff des sich selbst Vollendeten". 1785. In: Moritz, *Popularphilosophie, Reisen, Ästhetische Theorie*. Werke 2, Frankfurt 1997, p. 1035.

of whether architecture may, in the end, pass as art at all, or whether architecture-inherent potentials must not be comprehended as societal potentials, to be mastered anew without art or artificiality by all that are in touch with it.

That is the grand question.

In recent years in particular, architectural production was all mystified into being the deed of geniuses or, a tad less pretentiously, of stars.

But I surmise that architecture in the first place is not about art, short perhaps of a tiny remaining scrap. For Loos, art is confined to funeral and commemorative monuments. Other than that, in his view it is all about reasonably organizing societal work processes: "The house must please everybody. In contrast to the work of art, which does not have to please anybody. The artwork is the artist's private matter. The house is not. Art is produced without being wanted. The house meets a need. The work of art is responsible to nobody, the house to everybody. The artwork is revolutionary, the house is conservative. The work of art shows mankind new routes and is mindful of the future. The house minds the present. Man cherishes everything that serves his comfort. He loathes whatever dislodges him from his acquired and secure position, and upsets him. Therefore he loves the house and loathes art. Would then the house be bereft of any relation to art, and architecture no fellow of the arts? Quite so. [...] Since there are tasteful and tasteless buildings, people conclude that those are due to artists, and these to non-artists. But building tastefully carries no merit, just as not putting one's knife in one's mouth or brushing the teeth in the morning is of no merit."[18]

This is a rather nice debunking of architecture.

And then this sentence: "The single man is incapable of creating a form, hence so is the architect."[19]

The same point is being made by Goethe in some less well-known texts from a trip to Switzerland: "But perhaps the time is no more one of building churches and palaces. I for myself would recommend in both cases accommodating the communities in decent prayer halls and the families in comfortable and bright city and country houses."[20] This quite sums it up. Things are no more about churches and palaces. It is all about rooms that are appropriate to their usage—while at the same time about the loss of art.

"We poor artists of these last times."[21]

---

18   Adolf Loos, *Architektur*, in: Loos, *Sämtliche Schriften in zwei Bänden*, edited by Franz Glück, vol. I, Munich 1962, p. 315.
19   Ibid., p. 315.
20   Johann Wolfgang von Goethe, *Tagebücher* 1779, Weimar edition, III., vol. 1.
21   *Briefwechsel zwischen Schiller und Goethe*, edited by Emil Staiger, Frankfurt / Main 1977, p. 53f.

INDEXICAL MARKINGS OF THE TOPICS DISCUSSED

In a rather indexical and compact form, we would like to provide some coordinates of the issues discussed after Helmut Geisert's lecture "A few thoughts about the issue of appropriateness in architecture" at the Printed Physics Conference, and which forms the basis for his text in this chapter.

A first line of discussion was based around the topos of the cathedral as folk art. Geisert stated that the critical point with his parable was the issue of how the wishes, hopes and needs of the people could be connected with architecture today. The image of the cathedral described by Victor Hugo would stand for this connection. To speak of the end of the cathedral construction raises the question of how one can justify a new architecture. It was discussed to what extent this issue on the basis of the actual justification was actually a 19th-century issue, in the sense that they belonged to a time which—to quote from the discussion—"was all about providing leverage of history on an anthropological level in the most innovative ways". The core of the objections to the re-emergence of such fundamental questions today was formed by the criticism of the associated theories of need, and the inextricable connection of such theories with the assumption of anthropological constants.

The image of the cathedral is to be seen as a retroprojection not just in terms of its use in Geisert's parable, but also in Hugo's actual portrayal of the cathedral. It was described in the 19th century by Victor Hugo and referred to Viollet-le-Duc's restoration project for Notre Dame in Paris. On this basis, the discussions were generalized on—to quote from the discussion—"methodological states of emergency" historical perspectives find themselves confronted with. Various participants pointed out that the purpose of the historical connection in developing a contemporary would be misunderstood, if one were to see it as an authorization of a thesis by historical references. Much more it is to be regarded as an explication of a subject towards the future, with complete openness but with concrete offers for integration based on historical experience. Thus the suggestive vision of Geisert considers that the technological developments relating to digitalization, as elaborated for example in Hovestadt's lecture, could reverse the fragmentation of folk art accompanying the art of printing. One participant wished to sharpen this thesis and asked about where this suggested concept would lead if explicated from the form of an abstract suggestion, and what would happen with the claimed openness for the historical portrayal if the suggested parable was viewed in this way. In order to problematize such an accentuation, the image of the cathedral was compared to that of the Tower of Babel, in which everyone had easily and to some degree immediately understood each other while the tower was being constructed. The confusion arose first with regard to the question of why.

A second line of discussion arose from the question posed by the lecture regarding appropriateness. This was discussed in conjunction with the concept of living, which has always developed in various ways with the ideas of appropriateness and proportionality. The concept of living provided terminology for placing particular interests in relational conditions. An objection was raised in an argument against such a structural expansion of the concept of living, with living, at least as a relevant architectural term, only referring to the very short period of time with the advent of civil society. However, the discussions were in agreement that with the enlightenment, a shift had occurred in architecture from—to quote from the discussion—"building to living", which is key for the issue regarding appropriateness in architecture. While the issue of proportionality in clerical and aristocratic architecture can be justified with reference to a transcendent, there is a problem with justification regarding residential homes and apartments as fully secularized architecture. The move to a new architecture at the beginning of the 20th century aimed to free this justification history from being exclusively debated as an issue of aesthetics. In the wake of the Enlightenment, industrialization and the associated demographic changes, the house had become a prime architectural task. One main theme of modern architecture could be seen—to quote from the discussion—"in handing over the authority of aesthetics and beauty to a usage and practice". The problem with this development, some participants commented vehemently, is the naturalization of human behaviour that has taken place. The secularization of how to legitimize appropriateness for living space needs to claim the existence of anthropological constants, in order to have recourse to their systems of needs in a rational and objective way. It was also talked about in the discussions to what extent this scheme for a naturalization of the social and the cultural also stands behind the interest to turn architecture fully into science, and how through such a scheme of naturalization—meant to be a secularizing move—issues of appropriateness in fact still make reference to a transcendent, now in the form of social or anthropological constants.

A third line of discussion branched off from these thoughts and attempted to take an overall look at the range of implications associated with the reduced forms of radical functionalism, such as the general nodes of Konrad Wachsmann and Fritz Haller. These would encapsulate the proportional ideas of order in a relational concept of transformability, thereby relativizing the ideas of proportionality. The discussions focused on how to deal with the manifold implications that can be triggered in manifold ways on the basis of such reduced, radically functional formal nuclei or nodes. What does it mean to make important a development that has apparently just started to progress in one direction or the other simply by its explicit thematization, and the assessment of the development we thereby transport? The discussions claimed that a particular problem concerning the use of simplifications becomes apparent with regard to this question. Simplifications oscillate to a certain extent, according to necessity, between streamlined typification that aims simply to adapt to the system, and simplifications that aim to make transportable as much historical complexity as possible from specific specializations. Such simplifications are just as reductive in nature, but they seek to enable not the simplest but rather the most differentiated integration possible into a changed overall context. In the discussions, a perspective was pushed that proposed geometry as an exemplary valuable main protagonist in this tradition of typification and simplification. It was discussed about what happened with this traditional stabilizing role of geometry with regard to the emerging algebraization of geometry in the 19th century. The space within which one calculated with this other mathematical approach was a genuinely symbolic space that no longer involved measuring the earth and enabling it to be mapped proportionally, but this mathematics included the potential space for electricity and enabled resulting reference-less renderings. This symbolic potential space and the related algebraic mathematics, it was argued, underlie the most recent step of abstraction of the mechanically general node to form an information technological general node, the computer chip.

It was then discussed whether and how the computer chip is "existing" in a different way than mechanical nodes. This abstract statement was illustrated by the fact that chips no longer have to be screwed together, and no longer exist as pieces of metal that can easily be weighed. And they are printed in masses, yet the category of reproduction in Walter Benjamin's sense seems insufficient. Computer chips cannot be adequately described as the reproduction of an original, each of them is strictly speaking pre-specific in what it will come to have enabled, qualitatively, after being implemented somewhere. Even when we view this "informational" node as a material, as with the illustration, we can no longer measure it in kilograms and think in terms of magnitudes of 1,000s. We must measure it in grams and think in terms of magnitudes of millions. We are only used to this magnitudes from thinking about chemical elements perhaps, as raw material, or from the populations of species that concern biology. However, in terms of our intuitive thought and our imagination, understanding such magnitudes is not something for which we are well trained. For as long we as individuals have thought in terms of things that we can touch and handle as objects that this kind of intuition almost feels natural. However, some participants preferred to consider this escalation in terms of how to deal with magnitudes as non-purposeful, as it would not change—to quote from the discussion—"the thing in itself" with its sheer numbers. Yet in the case of things transformed on the atomic scale, in their physical properties, we are dealing with artefacts; what the "thing in itself" is supposed to mean in these cases lacks a "natural" reference. Thus, other participants were just as vehement to the contrary, so we must learn from these quantitative relationships a different qualitative way of converting and transferring with simplification than we could manage solely with the functional portrayals. A further contributor, somewhat impatient with the hesitation mentioned regarding the purposefulness of such abstract considerations, compared this situation allegorically with the difficulty of wishing to speak during the emergence of the printing press about the implications later raised by Hugo relating to the societal role of architecture. It was ironically asked whether we should effectively follow a messianic logic, and wait for new Victor Hugos before we may expect to be able to speak about the changes related to digitization of which we all know that they are happening now. The continued discussion highlighted ways regarding how one can learn to think today about the qualitative difference by creating contrast images. More controversially it was debated whether or not a sweeping and lower differentiation in detail was a strength of such contrast images. One line of argument in favour of the former held that the contrast images can only then provide an inventory of options with which one can learn to play in a differentiated manner without always applying the old schemes when details are avoided. This argumentation was in favour of creating as many contrast images as possible, all of them abstract and reduced. Other participants thought this unproductive, and preferred to look for the right and adequate ones immediately.

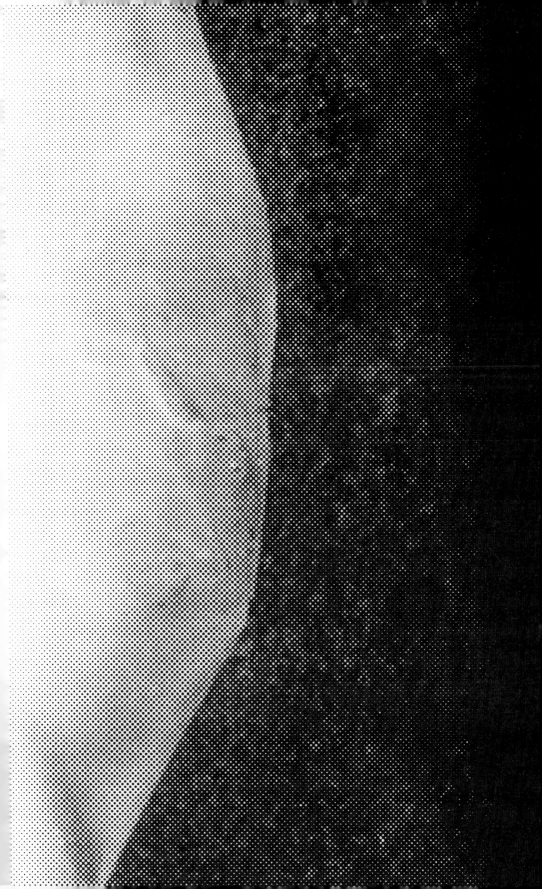

VI BRINGING AND POSITIONING: WAYS OF TECHNOLOGY? —APPROACHING HEIDEGGER'S THOUGHT ON TECHNOLOGY
HANS-DIETER BAHR

HANS-DIETER BAHR was professor of philosophy at the Free University in Berlin, at the University of Bremen, at the Faculty for Architecture in Milan, and until his retirement in 2000 at the University in Vienna. Among his publications are *Kritik der "Politischen Technologie"*, Europäische Verlagsanstalt, Frankfurt 1970; *Über den Umgang mit Maschinen*, Konkursbuch, Tübingen 1983; *Die Sprache des Gastes. Eine Metaethik*, Reclam, Leipzig 1994; *Zeit der Musse —Zeit der Musen*, Attempto-Verlag, Tübingen 2008; and *Die Anwesenheit des Gastes. Entwurf einer Xenosophie* (forthcoming 2011).

Martin Heidegger did not attempt to think the essence of modern technology anthropologically, on the basis of instrumental activity, but from "positioning" (*Stellen*).[1][↗P.218] Here the term "essence" does not refer to its type or genus, but to the way in which it comes to presence, and to what it reveals. This positioning of modern technology is still said to be "connatural" with the positioning (thesis, positio) of non-modern technology, in which the production remains bound to a nature (physis) that brings itself forth and thereby presents itself from itself; Heidegger claims, however, that the "main feature" of modern positioning has utterly changed.[2][↗P.218] What modern technology provokes, according to Heidegger, realizes itself as a positioning in the sense of a challenging-forth (*Herausfordern*) and as an "ordering" in the double sense of demanding and arranging, which are ways of positioning that link up with one another through allocating (*Bereitstellen*), delivering (*Zustellen*), and connecting (*Durchstellen*).

Until now Heidegger's concise remarks on modern technology have been scarcely taken up and developed in philosophical thought. The Heideggerian term "framing" (*Gestell*) should not be understood in the sense of a gearbox's frame, but as the gathering of ways of positioning—until now this term has not made its way into a broader philosophical terminology. Was the impertinence of this turn in thought too bold, which Heidegger equates with that of Plato, who rethought appearance (*eidos*) in the sense of the invisibility of the ideas?[3] What could be the difficulty in sufficiently understanding the essence of modern technology in terms of "positioning"? How, in general, would it be possible to understand the aspect of "demanding" within positioning, without thereby reducing it to the mere needs and desires of human subjects, which once more would define the concept of technology not in an ontological, but in a purely anthropological manner?

I will first outline how Heidegger understands non-modern technology, in which he, according to his example, detects a close relationship between natural, handcrafted, and artistic bringing-forth (*poiesis*), as well as social and cultic practice. Nevertheless, for him the character of this technology does not consist in its means-ends relationships, through

1   All italics are from the original and appear here following the English translation within parentheses. All Greek and Latin terms appear here as they do in the original, in parentheses and in quotations. Direct quotes from Heidegger have been translated within the text, followed by the German original in footnote with references to page numbers in the German editions (translators).
2   Martin Heidegger, *Einblick in das was ist.* Bremer Vorträge (1949), in: Complete Works, vol. 79, p. 67.
3   Martin Heidegger, *Die Technik und die Kehre.* Verlag Günther Neske, Pfullingen 1962, p. 19.

which humans partook in this bringing-forth, without having produced and mastered it on their own. Heidegger was not concerned with the question of how technologies can be useful, or perhaps, vice versa, how humans can be dominated by them, or even stay "neutral" towards their usage, but instead asks himself how and as what the character of technology shows itself from itself, that is, which way of revealing it is. I speak of "non-modern technology" because while Heidegger provides examples from pre-industrial technology, he thereby also indicates possible tendencies within modern technology opposed to positioning, which might—through bethinking the arts—be able to unfold their strength. For Heidegger claims that it is the "danger" of modern technology to conceal its own character as positioning, and furthermore to conceal this concealment, so that its true being would remain hidden. Heidegger contends that in order to understand its essence it is not sufficient to classify technology in a purely instrumental and anthropological manner, which is to say, as a means to ends and as human activity.[4]

Heidegger begins by remarking that "Wherever ends are pursued and means employed, wherever instrumentality reigns, there reigns causality."[5] In order to avoid the false impression that the ends relate to the means as causes relate to effects, it seems appropriate to me to make some preliminary comments. What concerns me here are not primarily the unintended or unpredictable side effects of technical activity, which might even in those cases, when they endanger the continued existence of humanity, exceed human control. Instead I want to focus on the structure of the means itself, which is actually not fixed with a determined (*bestimmten*) end—as the effect is by the cause—as was already demonstrated by Kant in his *Critique of Judgment*. Since particularly in applying (*Verwendung*) a mean, wherein the actual implementation suspends the characteristic relation between means and ends, the "means" are literally turned around. That is to say that they are finally turned into components (*Bestandteile*) of the process as a whole, which in their causal connection can still accord with the ends, but are no longer available as a means in a dispositional sense. Here we speak of "ends-oriented conditions", which present themselves again as available means only to a recollecting or anticipating reflection. The stone fitted into a wall, the wheel on a vehicle, and the computing program of a robot are no longer means in themselves but ends-oriented components. To the extent that they are reflected upon as means, there appears from the outset the possibility for them also to be the raw material of a sculptor, the missile of a warrior, or the material of a video artist. Therefore one of the characteristics of a means is to be

---

4   Ibid., p. 6.
5   Martin Heidegger, *The Question Concerning Technology and Other Essays*. Trans. William Lovitt. Harper & Row, New York 1977, p. 6. / "Wo Zwecke verfolgt, Mittel verwendet werden, wo das Instrumentale herrscht, da waltet Ursächlichkeit, Kausalität." ibid., p. 7.

available (*dis-ponibel*)—"dis-positioned" and "positioned-in-two", so to speak—out of which its maneuverability (*Wendigkeit*) is indicated: namely that a means is on hand for different ends even in those cases when its structure as a thing—whereby its possible applications are restricted—is not changed. In regard to many highly specialized means such as nuclear reactors, the signs of this maneuverability might be restricted to the possibilities of regulating the degree of their power, monitoring them, turning them on or off, or using them indirectly as the object of political disputes. Even with pieces of clothing, however, it is not the case that they are merely taken on or off, but wearing them is directed toward the most diverse purposes: according to weather conditions; as a sign of a sense of shame; in accordance with individual taste; in order to transmit erotic or aggressive messages; demonstrating one's affiliation to certain groups; and in order to demarcate oneself from others, etc. Beyond the use primarily assigned to them, they furthermore can be used as fetishes, as presents, as rags, as murder instruments, as models for a drawing, or—as in this case—as an example in a discourse. The majority of means such as containers, hand tools, vehicles, modern data processors, and others, demonstrate an immeasurable variety of possible usages. There are always degrees of availability depending on the conditions that restrict them. But only with the actual striking of the hammer on the nail, or when tapping a button in order to make a computer program run, does this striking or tapping operate as the cause's trigger and as an ends-oriented condition, but not as an available means in a dispositional sense.

Thus ends and means do not stand in a causal relationship to each other. It would be absurd to say that the idea of an end taken as a motive was the "cause" for choosing one means over another; this is so because the choice is based on ponderable reasons governing our behavior, and not on determined causes that only manifest themselves in their very realization. This might sound as if the means as a "capability" belong to a "realm of freedom" just as the causes as "compulsions" belong to a "realm of necessity". However, in that case one would have already overlooked the fact that the availability of the means is anything but under the control of a single person able to decide freely or choose. Rather it is from the outset a plurality (*Vielheit*) of human beings who—either independent from each other, or together with one another, or against one another—use those means. Furthermore, the latter one is positively received by the currently widely accepted opinion that competition and rivalry would promote not only deception, aggressive confrontations, and monopolies of power, but above all technical progress and economic prosperity. The means as a resource for the other person that he might use against me appears as something threatening to me. Be that as it may, the availability of means places us in opposition to the amount of their incalculable usages, such that societies have always attempted to minimize their own threatening dynamics through numberless proscribed uses

(*Gebrauchsvorschriften*). Customs, morals, practices, property rights, power, regulatory monopolies, norms of education, certain entitlements in a society structured by the division of labor, etc.: all of these have the general negative purpose of excluding countless, alternate, possible usages that in turn make necessary an industry of techniques for regulation and security, which then must be secured itself and so forth. We live in a social web of the unpredictable maneuverability of available means of the causally related, more or less ends-oriented conditions and normative mechanisms—as well as their implementation in technical, factual "constraints". It is very unlikely that anyone would still want to claim that this web as a whole is instrumentally controllable—subjectable to a human will—even if he or she were to act in the name of all others. It is rather the other way around, so that the representative single will is utilized by the social relationships. Hegel had already observed this with the help of Adam Smith, although today we are no longer able to recognize within these social relationships either an "invisible hand", or a "cunning of reason" (*Vernunft*).

In the description of non-modern technology Heidegger could not hold the view that the means-ends relationship is a causal one, because he was aware that this conception of causality is quite narrow. It was only with the mechanistic understanding of the natural sciences that the opinion was generally accepted that any kind of causality ultimately can only originate from one of the four forms of causality originally conceived by Aristotle, namely the "Wirkursache", or the *causa efficiens*. Alongside the example of the bringing-forth of an offering cup, Heidegger points—against this predominant concept of causality—to four interrelated ways of "thankfulness" or "indebtedness", by which he of course does not mean a moral or lawful obligation, but the founding of and accounting for a certain gathering. According to Heidegger, what is brought forth through the causes exceeds, from the outset, human activity and production. The offering cup owes to the silver its material (*causa materialis*); it appears in the form *eidos* of the cup-ness (*causa formalis*); and it is limited in advance within the realm of ordination and donation (*causa finalis*). Here its *telos* is not to be understood as an aim or end but as the full-filling of the bringing-forth. Finally, the silversmith is the one who is also to be thanked for the availability of the completed sacrificial device. He does not simply have an effect in the sense of the "*causa efficiens*" but rather gathers, in the sense of "*legein*" and "*logos*", the different ways of obliging, and it is through him that the bringing-forth and the upheld consistency of the offering cup has its origin.[6] Thus, on the one hand, the bringing-forth, the *poiesis*, is realized primarily through human activity, while on the other, this cannot be carried through by it alone, because

---

6   Ibid., p. 9.

humans cannot even attribute their own abilities purely to themselves. It is to nature as *physis*—as that which makes beings emerge from within itself—that the bringing-forth is also crucially indebted. Nature is even, in a higher sense, *poeisis* because it has not, like that which is brought forth by humankind, the origin of its bringing-forth in another, i.e. the craftsman or artist, but within itself. While the human being as that which understands the gathering of the other three causes in the sense of *logos* (*verstehe*) and as that which knows (*weiss*) how to conduct these in the sense of *episteme*, is through its constitutive thinking an essential contributor, it is not its God-like creator and producer. The causes bring together and usher the silver bowl into full presence within a realm of usage. It follows that the bringing-forth is a way of "revealing" the truth of that which presents itself to us as the essence of this technology, and that on which the possibility of all production is based:[7] "Technology comes to presence [*west*] in the realm where revealing and unconcealment take place, where *alētheia*, truth, happens."[8] In the sense of such unconcealing, *techné* is always also *episteme*—knowing as the ability to be receptive. And this does not only take place in the production and realization of "equipment", which is then on hand to be used and reused. Rather one might add that it is also in the manner of the usage and the way of dealing with the product that there is a bringing-forth, a "*pro ducere*", a fostering guidance, a caring for the arriving of that which is to be brought into the unconcealment, and for the giving of itself to be known as a distinct "destining-of-being" in this arrival.

In the *Bremer Vorträgen* Heidegger elaborated on this "destiny" (*Geschick*) insofar as he contrasts what modern technology provides with the essence of the thing (*Ding*). In this context "destiny" should not be confused with an unavoidable fate. What the "thing" gathers in a wine jug as thing is not simply the clay and the coming-forth of its form through the bringing-forth of the potter. Heidegger claims that since Plato, philosophy has not made the essence of a thing the starting point of its thought, but merely determined it as an object based on its appearance.[9] In contrast: "The pitcher is not a container because it was produced; rather, the pitcher had to be produced because it is this container."[10] In its ability to gather there lies a certain claim, to which the desire of humankind—by adopting and accomplishing this claim—is able to correspond. Neither the base nor the wall frames the vessel but the "emptiness", as Heidegger

---

7   Ibid., p. 12.
8   Ibid., p. 13. / "Die Technik west in dem Bereich, wo Entbergen und Unverborgenheit, wo *aletheia*, wo Wahrheit geschieht." ibid., p. 13.
9   Martin Heidegger, *Das Ding*, in: *Einblick in das was ist*, ibid., p. 7.
10 Martin Heidegger, *The Heidegger Reader*, ed. by Günter Figal, trans. Jerome Veith, Indiana University Press, Bloomington, IN 2009, p. 256. / "Der Krug ist nicht Gefäss, weil er hergestellt wurde, sondern der Krug musste hergestellt, weil er dieses Gefäss ist," *Das Ding*, ibid., p. 6.

observes, echoing a thought of Lao Tse. If one attempts to determine the jug in a purely physical manner, i.e. as that which it "actually" is, namely as already filled with air, one loses sight of this thing as a gathering. The emptiness, or perhaps it would be better to say, instead of defining it by its deficiency—the openness of the jug—gathers by receiving and containing what is poured into it.[11] It is pouring that determines it and which is a form of giving. Therefore the jar-ness of the jar is present in the bestowing of the pour, which might be a drink of water or wine. What belongs to earth and heaven dwells in the water or wine and the drink quenches the thirst of the mortal, but at times it can also be used for consecration. In this case the pouring is the drink administered to the immortal gods. Such pouring is not a mere pouring but an oblation and sacrifice:[12] "The gift of the pouring is a gift insofar as it houses earth and sky, the immortals and mortals."[13] The "mirror-game" of the fourfold (*Geviert*), as Heidegger puts it, is supposed to take place as "rather the inexplicable and unfathomable about the worlding of the world," as it gathers in the thing.[14] It is by sparing the thing as a thing that we humans dwell in the proximity of the world, and that we are the conditioned ones. Here Heidegger speaks of a "round dance of happening" (*Reigen des Ereignens*) through which the thing is liberated from its abstract isolation as an object.

I have briefly outlined what, according to Heidegger, is in danger of being forgotten due to the essence of modern technology. The essence of causality seems to be significantly changed. Through its limitation to effect and calculable success, so says Heidegger, not only a "neglect" of the thing comes to pass, but furthermore a concealing of the manner in which the nature of technology in the unconcealed presence of the brought-forth never ceases to be present. But how is it that we can understand "positioning" so that it nonetheless reveals the way modern technology comes to presence?

In *Die Technik und die Kehre* he initially briefly states: "And yet the revealing that holds sway throughout modern technology does not unfold into a bringing-forth in the sense of poiēsis. The revealing that rules in modern technology is a challenging [*Herausfordern*], which puts to nature the unreasonable demand that it supply energy that can be extracted and stored as such."[15] Since Heidegger merely indicates the manifold meanings

---

11 *Das Ding*, ibid., p. 10.
12 *Das Ding*, ibid., p. 12.
13 *Reader*, ibid., p. 260. / "Das Geschenk des Gusses ist Geschenk, sofern es Erde und Himmel, die Göttlichen und die Sterblichen verweilt," *Das Ding*, ibid., p. 12.
14 *Reader* ibid., p. 265. / "Unerklärbare und Unbegründbare des Weltens von Welt," *Das Ding*, ibid., p. 19.
15 *Reader*, ibid., p. 14. / "Das Entbergen, das die moderne Technik durchherrscht, entfaltet sich nun aber nicht in ein Her-vor-bringen im Sinne der *poiesis*. Das in der modernen Technik waltende Entbergen ist ein Herausfordern, das an die Natur das Ansinnen stellt, Energie zu liefern, die als solche herausgefördert und gespeichert werden kann." Martin Heidegger, *Die Technik und die Kehre*, ibid., p. 14.

of the term "positioning", which are gathered in "framing", I will begin by attempting to outline some of those basic meanings that assemble here.[16] A thing can be at any position (*Stelle*). In contrast to the determined place (*Ort*) that a thing like a tree, a house, a thought, in its context occupies as its own, such a position randomly substitutes the position of another position—like a stone at a pile of debris, or a leaf in a pile—that might also be positioned at any other point, just as it is irrelevant which particular spot similar things occupy in a warehouse. A change has already occurred when something is positioned (*gestellt*) in a position, when for example a vase is positioned on (*auf*) the table, or a shoe in (*in*) the closet, or a stool under (*unter*) the table, or when someone positions himself at (*an*) the window in order to look out for something. In spatial arrangements (*Anordnungen*) like these a random substitution is no longer possible. Frequently the positioning occurs as a setting-upright: one positions a stool's legs or gets on one's feet, sets something up, or sets books in order that are lying around. Positioning one thing above another can be an expression of preferring or neglecting something. Someone can in general be good or well-off in regard to his living conditions (*gut oder schlecht gestellt sein*). In our social dealings people pursue activities, obtain a certain position and attitude (*Einstellung*), and occupy a higher or lower position. All these ways of positioning are no longer about random positioning but about hierarchies or arrangements of equality. Items, processes, or activities are brought into the required position with one another. For example, one adjusts (*stellt*) a watch according to a normative form of movement; one turns on a machine and adjusts (*stellt*) it to a certain pace so that it functions adequately; and one adjusts (*stellt*) human beings and hires them so that they fulfill certain tasks. Here "positioning" already means a composition (*Erstellen*) so that something definitive is accomplished. The doctor gives (*stellt*) a diagnosis; the meteorologist forecasts (*eine Prognose stellen*); the salesperson writes (*erstellt*) a bill; and the state sets a budget. Certain items or accomplishments are produced (*hergestellt*) in a way so that in their having-been-produced they are at the same time readily positioned, that is to say, completed and positioned in order to be delivered (*zugestellt*). They are partly available for other applications because their production (*Herstellung*) has already been ordered (*bestellt*). In these cases such an "ordering" is related to an assignment and demand (*Anforderung*) to produce something, to deliver it, or to report to somebody. A demanding ordering (*Bestellen*) can also be directed to oneself, for

---

16   Throughout the remainder of this paragraph the majority of the terms used and discussed contain some variation on the word *stellen*. English translations of these terms cannot convey its ubiquity in everyday, spoken German. The author's intention is to illustrate the manner in which *stellen*—meaning variously to "position," "place," "put," "order," etc.—pervade modern language and practice. All quotations in parentheses provide the variation on stellen as it appears in the original (translators).

example, when one has to till (*bestellen*) a field or when certain aspects of one's life are in disorder (*nicht gut bestellt*). We also speak of a demand when we test someone (*auf die Probe zu stellen*), when we face a difficult task, problem, a quarrel, or an uncomfortable truth or duty; in this case a courageous attitude (*Einstellung*) might be necessary in order to be able to deal with it instead of avoiding it. In such cases a particular position is occupied.

Thus, such a positioning entails making something stop, react, or even forcing someone to change, maintain, or defend his or her own position against its distinctive tendency or inclination. Therein lies a challenging (*Herausfordern*), through which something is drawn out that would not have been actualized by itself even though it has the potential for this. This is because one cannot challenge unless there is the potential for this challenge in the thing itself. One can obstruct somebody or block their way; one can bring someone before the law; hand someone the draft notice for military service; one can force someone to position himself and thereby attempt to acquire and maintain a particular position oneself, so that things might come to a "war of positions"; and when the captured one (*der Gestellte*) tries to escape one can chase (*stellen*) him. In case something attempts to avoid this challenging positioning one can set a trap for it. In an essay falsely attributed to Aristotle from around 200 AD entitled "*Mechane*" someone even spoke of a "finessing of nature", since it is through the lever arm—which is lengthened to one side of the pivot— that heavy weights of any kind can be moved by minor human or animal powers. This is "cunning", because where there is no economy of time it does not matter that at the same time the route of transportation gets longer. This positioning produces something out of nature that under certain circumstances it would not have produced on its own. Every challenge, particularly when challenging an enemy to fight, entails a provocation. One forces the other to acquire and display certain behaviors which he would not have acquired from his own initiative, though he nevertheless had the capacity for them. In this sense every challenge is at the same time an encouragement and a clearing-up of that which would not have shown itself on its own account.

All these ways of positioning are not concealed from us but rather the positioning and the positioned thing are more or less positioned in front of our eyes (*vor Augen gestellt*): they are presented either in an accusing manner when pilloried, or in an enticing manner. Therefore they can be displayed, exhibited, or presented by us, but we cannot be sure whether what is positioned in this way is not distorted (*entstellt*), or as it is the case with a trap, if important possibilities are blocked.

Because we are fully adapted to the doctrine of the natural sciences, according to which nature does not pursue aims, but at the most approaches an end in an entropic manner, we do not want to assume that natural processes even in the form of closed-technical sequences would "position" human beings in the manner of positioning the Other. This suggests, however, that each and every positioning-challenging seems to be caused only by human desire and doing. And yet still we claim that natural states and processes as well as technical inventories and sequences challenge us and demand something of us. What then can be meant with the claim that positioning is the essence of technology? First of all I want to provide two of Heidegger's examples that make clear how positioning transcends the actions of individual persons and their views on means, and how it involves itself in this action as something that is ordered and engaged: Heidegger claims that cultivating the land is no longer a rustic cultivation of the soil, but is instead part of a motorized food industry, or it consists, for example, in the cultivation of a coal reservoir: "Yet toward what, for instance, is the coal geared that the coal dispensary is made to surrender? It is not placed forth as the pitcher on the table. Just as the ground is made to supply the coal, the latter is challenged for heat; this is already made to supply steam, whose pressure drives the machinery that keeps a factory in production; a factory that is made to supply machines that make tools with which, in turn, machines are brought to and kept in working order. One forcing [*Stellen*] challenges the other...."[17] From this "chain of positioning and ordering" it follows that a positioned thing is setup "...not in order to come to presence, but rather in order to be forced to supply something else."[18] Thus the nature of positioning itself entails a steady further-positioning and connecting to the next ordering: "The forester, for example, who measures the felled timber in the forest and appears to follow the paths in the same way as his grandfather, is today put to use by the timber industry. Whether he knows it or not, he is in his own way an inventory-piece of the cellulose-inventory and its assailability for paper, which is delivered to newspapers and magazines, which then stand to be consumed by the public."[19] Heidegger speaks of the "gear", of the "activity", of the "circulation of the ordering of the orderable into the ordering", which cannot be reduced to the manipulations of humankind but instead constitutes the being of modern technology: within the food industry as well as in the "fabrication of

17 *Reader*, ibid., p. 270 / "Wohin wird nun aber z.B. die im Kohlerevier gestellte Kohle gestellt? Sie wird nicht hingestellt wie der Krug auf den Tisch. Die Kohle wird, gleichwie der Erdboden auf Kohle, ihrerseits gestellt, d.h. herausgefordert auf Hitze; diese ist schon daraufhin gestellt, Dampf zu stellen, dessen Druck das Getriebe treibt, das eine Fabrik in Betrieb hält, die daraufhin gestellt ist, Maschinen zu stellen, die Werkzeuge herstellen, durch die wiederum Maschinen in Stand gestellt und gehalten werden.— Ein Stellen fordert das andere heraus." Martin Heidegger, *Das Ge-Stell*, ibid., p. 28.
18 *Reader*, ibid., p. 271 / "nicht um anzuwesen, sondern um gestellt zu werden und zwar einzig daraufhin, anderes zu stellen." ibid. p. 28.
19 *Reader*, ibid., p. 277 / "Der Forstwart z.B., der im Wald das geschlagene Holz vermisst und dem Anschein nach noch wie sein Grossvater in der gleichen Weise die selben Wege geht, ist heute von der Holzverwertungsindustrie gestellt. Er ist, ob er es weiss oder nicht, in seiner Weise Bestand-Stück des Zellulosebestandes und dessen Bestellbarkeit für das Papier, das den Zeitungen und illustrierten Magazinen zugestellt wird, die über die Öffentlichkeit daraufhin stellen, verschlungen zu werden." ibid., pp. 37–38.

corpses in gas chambers and extermination camps". Therefore the circular motion of ordering is not sustained just by greed for loot and profit, but is founded only in itself.[20] It is only as a positioning and positioned thing, as ordered thing and ordering, that we assign to Being a random place where it is only put into standing and "reserve", in order to provoke another positioning and ordering: "It aims for nothing—for the besetting does not produce anything that could have a presence for itself outside of the forcing. What is beset is always already and only ever forced to supply and bring something else to success as its result. The chain of besetting does not aim for anything; instead, it just enters its cycle. Only in this does the assailable [*Bestellbare*] have its persistence [*Bestand*]."[21] Thus it is not only the totality of apparatuses and machines that are part of the sphere of modern technology, but as well the techniques of economics—the challenge to achieve maximum output with minimal expense. Heidegger admittedly does not directly refer to the function of money as capital, which more than anything else represents the character of demanding and challenging, but his talk of the "circulation" of positioning allows us to conclude that he meant to include them. Having experienced the wars and mass murders of the 20th century probably prevented him from declaring capital to be the crucial medium of "positioning". Thus what is challenging-positioning as a whole are no longer, as it was said in the beginning, only "energies", but a positioning that essentially challenges a further positioning, which again is made available for further ordering.

According to Heidegger it is clearly not enough simply to speak of the functional mechanisms of a highly technological, differentiated, industrial society, which has—when it threatens to collapse—to care about its self-preservation and the increased efficiency of its systems and subsystems, upon which the usages and developments of its technologies are determined. The basic character of positioning is rather the demanding and challenging of what does not, like that which is to be brought forth, finally in its presence come to an arrival and rest as a finite product: since in each case the positioned thing, challenging, and challenged thing, so to speak, circle around—as something that in its turn has to be further constituted—with that which is other to itself. In the self-preservation or self-improvement of a highly technological social system, e.g. the one Niklas Luhman has in mind, one at least would be able to detect a meaning as final end.[22] Positioning, however, seems to be endlessly circling in

---

20 Ibid., p. 27, p. 29, and p. 33.
21 *Reader*, ibid., p. 271 / "Die Kette solchen Bestellens läuft auf nichts hinaus; denn das Bestellen stellt nichts her, was ausserhalb des Stellens ein Anwesen für sich haben könnte. Das Be-Stellen ist immer schon und immer nur daraufhin hingestellt, ein Anderes als seine Folge in den Erfolg zu stellen. Die Kette des Bestellens läuft auf nichts hinaus; sie geht vielmehr nur in ihren Kreisgang hinein. Nur in ihm hat das Bestellbare seinen Bestand". ibid., pp. 28–29.
22 cf. Niklas Luhmann, *Zweckbegriff und Systemrationalität*. Suhrkamp, Frankfurt a.M. 1968.

itself—and thereby reminds us of Hegel's term "bad infinity"—and thus the issue of whether it has a meaning or not seems to lie outside the question. And yet the positioning in its dynamic of challenging all but causes a mechanistically represented repetition of the same. It is, as Heidegger thinks, "universal", and everything that is belongs to its reserve and is a component of its reserve in its arbitrary substitutability.[23]

How are we to understand that a positioning-challenge not only challenges Being, but more so exposes itself as the essence of technology, and at the same time hides this, its very truth in the idea that it is purely the product of humankind and will serve in the name of its continued existence? The universal positioning itself, insofar as it aims for nothing, is also no longer partly determined by a *causa finalis*. Despite the cult of great technical achievements nothing is full-filled within them any more—nothing attains presence within them as something that is fulfilled. From the outset they merely function as parts of what follows, as elements in a series. In what sense, however, can positioning still be determined as essentially "challenging", if the challenging as a whole is neither aiming for an end, nor is it a mechanistically coerced causing and happening? Heidegger emphasized that the circulation of positioning is not to be understood on the basis of the machine but the other way around: "The rotation of the machine is forced—i.e. challenged and made constant—in the circulation that lies in the mechanism, the essential character of the *Ge-Stell*."[24] The machine exists only insofar as it is going, and it runs in the mechanism of operation, which operates the activities of ordering the orderable.[25] Heidegger, who also opposed a demonization of technology as power—which assumes an independent existence opposed to the interests of humankind—nevertheless refers to it as a way of being, which reveals and at the same time conceals itself as something that is "raging and being-carried-away" and pulling [*Geraff*].[26] This probably means that the chain of positioning furthermore takes place with a snatching rapidity that ultimately no longer allows the positioned to be able to come to presence with the exhibited—if it were just to be presented as an object—not to speak of

---

23   Martin Heidegger, *Das Ge-Stell*, ibid., p. 40.
24   *Reader*, ibid., p. 274 / "Die Rotation der Maschine ist gestellt, d.h. herausgefordert und beständigt in der Zirkulation, die im Getriebe, dem Wesenscharakter des Ge-Stells, beruht." ibid., p. 34.
25   See ibid., p. 35. / (*Reader* 275). The view in the middle ages and continuing into the modern period was that the essence of the machine consists in "transmission" (*Transmission*). Over and above the mere mechanistic interpretation of trans-mission, namely as a transmission of powers to their function as tools, one had in mind the "transmitting-over" in which at the same time some sort of "dedicating" was at work. Thus the trans-mission of the traditional mills was still in contact with the "bringing" of the bringing-forth. See my book: *Über den Eingang mit Maschinen*, Tübingen, 1983.
26   Ibid., / "pulling" (*Reader* 273). See: Günther Anders, *Die Antiquiertheit des Menschen* vol. II, *Über die Zerstörung des Lebens im Zeitalter der dritten industriellen Revolution*, Munich 1980. Anders strips the "positioning" of technology by way of the "imaginable" from any kind of concealment.

the possibility of being perceived as a thing gathering the world. In any case, most of the technical procedures are no longer visible, although some of them are gaining more and more importance for audible experience. The "rapidity", the growing speed of positioning and ordering, are part of the concealing character of the nature of technology.

Next, Heidegger responds to the possible objection that technology assumes powers and substances of nature that escape this universal framing. According to Heidegger the natural sciences actually had long included nature in the realm of technical inventory. In regard to physics, natural powers are only accessible in their effect, and it is only in this context that power shows or displays the calculability of its magnitude: "The coming-to-presence of nature consists in effectiveness. In it, nature can immediately bring something into place, i.e. make it occur. Force is that which brings one thing to effect something else out of it in a surveyable fashion. The forces of nature are presented by physics in the sense of forcing through which the *Ge-Stell* secures what is present. Nature stands over against technology only in such a way that it persists as a system of ordering successes out of the effective that is secured."[27] This "orderability of success" also refers to the scientific determination of substances of nature as "matter". Their main feature is determined as inertia, as a constancy in movement. Inertia is the resistance against changes in movement. Thus, substances in turn are presented by way of power: "For physics, nature is the inventory of energy and matter...The pre-calculability of natural processes, decisive for all natural-scientific presentation, is the presentational assailability of nature as the inventory of success...In the age of technology, nature is no limit to technology. Nature is much more the basic inventory-piece of the technical inventory—and nothing else."[28]

---

27 Reader, ibid., p. 280 / "Das Anwesen der Natur besteht in der Wirksamkeit. In ihr kann die Natur auf der Stelle etwas zur Stelle bringen, d.h. erfolgen lassen. – Die Kraft ist jenes, was etwas daraufhin stellt, dass aus ihm anderes in einer übersehbaren Weise erfolge. Die Naturkräfte sind durch die Physik im Sinne des Stellens vorgestellt, durch welches das Ge-Stell das Anwesende stellt. Die Natur steht der Technik so und nur so gegenüber, dass Natur als ein System des Bestellens von Erfolgen aus dem gestellten Wirksamen besteht." ibid., p. 41.

28 Reader, ibid., p. 280-1 / "Für die Physik ist die Natur der Bestand von Energie und Materie... Die für alles naturwissenschaftliche Vorstellen massgebende Vorausberechenbarkeit der Naturabläufe ist die vorstellungsmässige Bestellbarkeit der Natur als Bestand eines Erfolgens... Im Weltalter der Technik ist die Natur keine Grenze der Technik. Die Natur ist da vielmehr das Grundbestandstück des technischen Bestandes – und nichts ausserdem." (ibid. Robert Spaeman makes a similar point: "Auch die gänzliche Zerstörung dieser Biosphäre auf diesem Planeten durch den Menschen kann ja als naturgeschichtliche Transformation verstanden werden. Eine Müllhalde ist – so gesehen – nicht unnatürlicher als eine Bergquelle... Der vollendete Technizismus ist so zugleich vollendeter Naturalismus." (Natur, in: *Philosophische Essays*, Stuttgart 1983, p. 36.). "Also the entire destruction of this biosphere on this planet through man can be understood as a natural-historical transformation. A landfill is—from this perspective—no more unnatural than a mountain spring...complete technicism is at the same time complete naturalism" (translators). Leaving nature only occurs when nature is remembered as herself.

If, as Heidegger emphasizes, man does not belong, like a machine, to this ordering that circulates in itself, how is he included in the circulation of positioning and ordering? It seems to me that this inclusion is possible only if the circulation of positioning as a whole does not also proceed as a mechanical effect. It may be that it inspires the appearance of a total causation of one positioning by the other, but it nevertheless does not follow an unbroken influence. According to Heidegger, man is a being that understands itself as something affected, approached, and addressed by things, occurrences, and its own experiences. In its needs there is as little a compulsion without beginning and end as there is in its will and decision an absolute source from within themselves. Its own desire, rather, is the way in which human beings are able to correspond (*entsprechen*) by thinking to that which addresses them. It is only in and through their own desire that they understand what is challenging them and what is asked of them, and what they in turn position and order in a challenging way. To me the difficulty here seems to be to understand positioning and ordering in a challenging way as happenings in Being, which, on the one hand, do not exist without human thinking, desiring, and doing, but rather only here reveals its inventory, while on the other, it transcends these and seems to be "sent" from somewhere else, like a permanently raised claim and demand. This destiny in the technological age of modernity appears to humankind to happen in a way in which their desire and being-demanded correspond to the technological challenges, which come too close to them, and are eventually too close (*zunächst*) to be seen in their questionability. It is in this and not in a naive renunciation of technology that the actual danger lies for Heidegger.

In *Being and Time*, Heidegger still claimed that all "equipment" aims to "eliminate" distance, so that proximity is established. This is not, however, to be understood in a measurable sense or manner, because the person who calls is, so to speak, closer to oneself than the receiver in one's hand.[29] In the care-structure of Being lies an essential tendency to draw closer: "the hasty elimination of all distances does not bring about nearness; for nearness does not consist in a small measure of distance."[30] What is this proximity when its absence means that distance also stays away? The elimination of distance and the approachability of anything and anyone at any time no longer brings nearness, but rather washes everything together into a uniformity without distance. "Is the convergence into the distance-less not more uncanny than everything blowing apart?"[31] In the clueless anxiety about the hydrogen bomb, which would

29 Martin Heidegger, *Sein und Zeit* § 23, ibid., p. 107.
30 *Reader*, ibid., p. 254 / "...das hastige Beseitigen aller Entfernungen bringt keine Nähe; denn Nähe besteht nicht im geringen Mass der Entfernung." M. Heidegger, *Einblick in das was ist*, ibid., p. 3.
31 *Reader*, ibid., p. 254 / "Ist das Zusammenrücken in das Abstandlose nicht noch unheimlicher als ein Auseinanderplatzen von allem?" ibid., p. 4.

be sufficient to eliminate life on earth altogether, it remains concealed that the horrific *has* already taken place, namely that all that is, is put outside of that which used to be its essence.[32]

Despite all the marks that the horror of the Second World War left on Heidegger's thoughts on technology, one should not in any way understand such sentences in a pessimistic manner. Indeed there is a concealment in the rapidness of technological challenging in which Heidegger sees the great danger, namely that it could withdraw itself from thinking and thereby from a revelation of the essence of this technology: this is to say that modern technology is able to conceal and reposition its own revealing and positioning in such a manner, that what is accomplished as an epoch of Being, namely in the way of Being of modern technology, can be taken for an act of man that he himself knows how to control and secure; herein we may directly sense a problem of language: not only in the fact that language can be reduced to a calculable amount of signals and thus to an orderable inventory of information, which are supposed to create images within us, but also in the fact that it enables us almost only to talk of positioning as if it were purely an act of humankind.

And yet perhaps in our reflection on of the essence of modern technology itself there is already a turn announced that can be a rescue from this danger in the midst of that-which-has-no-distance, from that which controls the nexus in which one positioning chases the other. This is so because the essence of modern technology also still lies in a way of revealing the truth, although it still conceals the retreat into a primordial revealing. Heidegger finally suggests that the inexorable ordering, and the restraint of what rescues, form a "constellation", whose secret lies perhaps in art.

> Works cited for translation purposes:
> – Heidegger, Martin. *The Heidegger Reader.* Ed. Günter Figal. Trans. Jerome Veith. Indiana University Press, Bloomington, IN 2009.
> – Heidegger, Martin. *The Question Concerning Technology and Other Essays.* Trans. William Lovitt. Harper & Row Publishers, New York 1977.

INDEXICAL MARKINGS OF THE TOPICS DISCUSSED

In a rather indexical and compact form, we would like to provide some coordinates of the issues discussed after Hans-Dieter Bahr's lecture "Bringing

---

[32] Ibid., p. 4.

and Positioning: Ways of Technology? Approaching Heidegger's Thought on Technology" at the Printed Physics Conference.

A first discussion thread developed around the cultural-historical meaning of ontological language games that depart from the idea of an initial positioning, as in *ponere, positio* or *thesis*. These language games reveal a certain feature that has played a key role in determining Western thinking at large, and which stands also behind Heidegger's characterization on technology as a circulatory system of positioning processes (*Kreisgang des Stellens*) as introduced by Bahr. There was a discussion of the difficulties that arise from the circumstances that the current technology not only creates and facilitates this positioning processes, but also presides in a broader sense over what is thought to be realistically feasible. Heidegger's figure of the cycle (*Kreisgang*) refers to this. Seminal to the discussion here was the fact that Heidegger's view of this management and facilitator role played by technology seems to guide our thinking about technology itself in a situation where the distance our thinking needs to maintain from the phenomena thought is getting lost. There was a strong opinion that, viewed in terms of the historical reception of Heidegger's thought, his characterization of technology has at least contributed to the fact that the earlier complexity and sophistication of the concepts of *techné* and *ars* are increasingly being lost in favor of much more vague terms such as that of art by will (*Kunst-Wollen*) or the even more activist terms calling for a politicization of the aesthetic. The art terms of *ars* and of *techné* referred once to much more than an aesthetic capacity, and, very similar to the term *poiesis*, were derived from and adjusted within the capacities of thought as a whole. Heidegger would undoubtedly have known this, was the consensus in the discussion. The critique was raised, however, that with the introduction of this assumption of an unavoidable lack of distance, Heidegger helped to dissolve a conceptually sophisticated and general, experience-based corpus of concepts that had existed in the traditional thought about *poiesis*. The objection is that this corpus of concepts concerning the capacity for thought has always had to wring the space needed for critical distance in the most difficult and arduous manner—as quoted from the discussion—"from the immediacy and immediate practical necessities". In particular it has been criticized that Heidegger breaks with this tradition and disperses it without sufficient attention on how to avoid falling behind the sophistication of the conceptual achievements he wants to break with. This argument was discussed with a great deal of disagreement, and even its legitimacy as an objection was questioned. A general difficulty emerged in this discussion, however, that concerns—quoted from the discussion—the legitimacy of wanting to describe something "in essence" (*der Sache nach*) or how, to a certain extent, as was objected, possible consequences such a fundamental characterization always has, had to be anticipatorily incorporated to a certain degree.

These discussions took another direction in the discussion thread regarding the topicality of Heideggerian thinking for our current situation. The discussion considered on the one hand, to what extent the solar cells do not precisely break out of this circulatory system of positioning processes (*Kreisgang des Stellens*), as a chain between producing, facilitating and managing. Because they make an energy potential available that had not prior been integrated into the ecosystem of the planet. Viewed in this way, we, as humans able to live on energy that we can obtain using photovoltaics, appear to be taken out of the cycle of nature as we know it. This line of discussion focused on clarifying the difficulties involved when referring to the concept of openness. It was subsequently continued by discussing whether a breaking off, or the possibility for breaking off from the radical chain of serialization can already be found in Heidegger's thinking about technology, for example by taking into account rather unintended side effects, cross-linking, active association or similar phenomena. Such phenomena have a great deal to do with the networks in which Heidegger's positioning processes are managed today. Various opinions suggested the importance of a probabilistic, non-causative view, as this would be more adequate for information technology. It was also discussed to what extent the processes that are currently administratively positioned—unlike during the post-war period in Germany in which Heidegger was writing, where all of the economic and political circumstances were oriented towards rapid growth—are not much less compelling overall. In addition the miniaturization, the cost-effective ubiquity, the sheer quantity and networking of today's information technology would change the processes described by Heidegger. It was discussed how his processes of positionings have gained a procedural und differential quality as a result of these changes, how they have become softer and more plastic today. These discussions were controversial. It was argued that the concepts of mechanical control and chain-linking do not appear to fit the phenomena of the present. But the perspective was highlighted that today, there is no factual necessity anymore for ascribing technology the presiding role with regard to governing the management of Heidegger's circular system of positioning processes. It was even suggested that the failure to recognize these changes is distracting from the unusual unrestraint and casualness of the current technology, and thereby locks the discussion of the various globalization debates in a fundamental bottleneck. A provocative formulation suggested that technology today is no longer valuable in an economic sense, but rather—as a politically manageable dispositive—is being made programmatically valuable, and that this occurs precisely through the postulation of such cycles (*Kreisgänge*), for example, in the sustainability debates from life cycle assessment to ecological footprint and its normalized—i.e. mechanized—econometrics for calculating planetary capacities.

IMAGE REFERENCES

All websites were last accessed on 3 August 2011. We have taken great care to identify all rights owners. In the unlikely event that someone has been overlooked, we would kindly ask that person to contact the publisher.
FIG. 1: Image Source: http://en.wikipedia.org/wiki/Plimpton_322 · **FIG. 2**: © Mathematical Association of America, Original from the collections of the Rare Book & Manuscript Library of Columbia University. http://mathdl.maa.org/mathDL/46/?pa=content&sa=view Document&nodeId=2591&bodyId=3434 · **FIG. 3**: ©Photographer: William A. Casselman, Museum catalogue of Museum of Archaeology and Anthropology at the University of Pennsylvania. http://www.math.ubc.ca/~cass/Euclid/papyrus · **FIG. 4**: ©Photographer: Robert Schmidt http://www.fisherman.is/blog/robert/template_permalink.asp?id=128 · **FIG. 5**: Producer of the foil: Holst Centre Eindhoven http://www.holstcentre.com/NewsPress.aspx Image source: http://www.gruponeva.es/blog/categorias/iluminacion/73.html · **FIG. 6**: Image source: http://www.exposolar.org/2010/eng/center/contents.asp?idx=88&page=1&search=&searchstring=&news_type=C · **FIG. 7**: Image source: http://3.bp.blogspot.com/_yxoiTroipWY/TLdyscmLmtI/AAAAAAAAAQ8/seg4D92s6Bg/s1600/Epaper-Fujitsu-742628.jpg · **FIG. 8**: ©Johannes S. Sistermanns, image source: http://www.khm.de/kmw/klanglabor/?m=201012 · **FIG. 9**: ©Ludger Hovestadt · **FIG. 10**: Image source: http://freakymartin.com/2008/03/29/miniplanets-38-photos · **FIG. 11**: Film still from Peter Greenaway's *The Draughtsman's Contract*, 1982 · **FIG. 12**: ©Stefan Tuchila, image source: www.archiphotos.com · **FIG. 13**: Photographer: Michael Schönholzer, image source: http://bikevisions.blogg.de · **FIG. 14**: Film still from Peter Greenaway's *The Draughtsman's Contract*, 1982 · **FIG. 15**: Image source: http://my.opera.com/tss101g/albums/showpic.dml?album=3476752&picture=57705382 · **FIG. 16**: ©Ludger Hovestadt · **FIG. 17**: Courtesy of Texas Instruments http://www.tf.uni-kiel.de/matwis/amat/semitech_en/kap_7/illustr/i7_1_2.html · **FIG. 18**: ©Ludger Hovestadt · **FIG. 19**: Image source: http://www.colepapers.net/tcp.archive/cole_papers_00/TCP_00_02/e-ink-n-paper.html · **FIG. 20**: Image source: http://www.pulsarwallpapers.com/r_club_music_wallpapers_24_vinyl_groove_macro_music_62690.html · **FIG. 21**: ©Ludger Hovestadt · **FIG. 22**: ©Stefan Kolb · **FIG. 23**: Unfortunately, the image source cannot be reconstructed any more · **FIG. 24**: Image source: http://genekeyes.com · **FIG. 25**: ©Ludger Hovestadt · **FIG. 26**: Image source: http://cache.gawker.com/assets/images/4/2010/01/wvga_and_original_ti_dlp_pico_chip.jpg · **FIG. 27**: ©Gramazio & Kohler, image source: http://www.robot-rent.nl/cms/index.php?option=com_content&view=article&id=17&Itemid=8 · **FIG. 28**: ©Nicolas Kruse, image source: http://de.wikipedia.org/wiki/Schrittmotor · **FIG. 29**: Image source: http://www.ati.surrey.ac.uk/news/open_day · **FIG. 30**: ©CERN, image source: http://cdsweb.cern.ch/record/910381#02 · **FIG. 31**: Image source: http://www.eteda.org/the_region/regional_profile/cocke_county.aspx · **FIG. 32**: ©ag4 | mediatecture company®, Cologne · **FIG. 33**: ©Serych at cs.wikipedia, image source: http://de.wikipedia.org/wiki/Datei:Quarzhalter.JPG · **FIG. 34**: ©Zaha Hadid Architects, image source: http://www.topboxdesign.com/wp-content/uploads/2010/07/Venice-Biennale-Design-by-Zaha-Hadid-Architects.jpg · **FIG. 35**: ©Frank Gehry, image source: http://commons.wikimedia.org/wiki/File:Duesseldorf_1960.JPG?uselang=de · **FIG. 36**: ©Ludger Hovestadt · **FIG. 37**: ©Ludger Hovestadt · **FIG. 38**: after the German edition of Michel Serres, "Gnomon: Die Anfänge der Geometrie in Griechenland", in: Elemente einer Geschichte der Wissenschaft, Suhrkamp, Frankfurt a/M 2002. Image source: http://earthfromabove.blogspot.com/2008/01/egyptian-pyramids-giza-egypt.html · **FIG. 39**: Unfortunately, the image source cannot be reconstructed any more · **FIG. 40**: ©Ludger Hovestadt · **FIG. 41**: Unfortunately, the image source cannot be reconstructed any more · **FIG. 42**: Image source: http://upload.wikimedia.org/wikipedia/commons/a/a8/NASA-Apollo8-Dec24-Earthrise.jpg · **FIG. 43**: Image source: http://i237.photobucket.com/albums/ff156/already_dead_photos/Internet_map_1024.jpg · **FIG. 44**: ©NASA, image source: http://www.jsc.nasa.gov · **FIG. 45**: NASA weather satellite, Hurricane Katrina on 28 August 2005. · **FIG. 46**: Image source: http://whatistug.deviantart.com/art/Night-Time-in-Tokyo-93421523?moodonly=24 · **FIG. 47**: ©Ludger Hovestadt · **FIG. 48**: Photographer: Tim McNerney, based on Intel archival materials, image source: http://4004.com · **FIG. 49**: ©Ludger Hovestadt · **FIG. 50**: Image source: http://www.mikrocontroller.net/topic/97237 · **FIG. 51**: Image source: http://www.isn.ucsd.edu/courses/490/2000/ADC/image002.jpg · **FIG. 52**: ©Infineon, unfortunately, the image source cannot be reconstructed any more · **FIG. 53**: ©Ludger Hovestadt · **FIG. 54**: ©Ludger Hovestadt · **FIG. 55**: Image source: http://commons.wikimedia.org/wiki/File:Flegel_-_Stilleben_mit_Äpfeln.jpg · **FIG. 56**: ©Frank Boller, image source: http://www.fboller.de/fotos/wallpaper/index.html · **FIG. 57**: Vincent van Gogh, Still Life with apple-basket, Nuenen, September 1885.